Develop[...] Nu[...]

MEASURES, SHAPE AN[...] E

ACTIVITIES FOR THE DAILY MATHS LESSON

Hilary Koll and Steve Mills

A & C BLACK

Contents

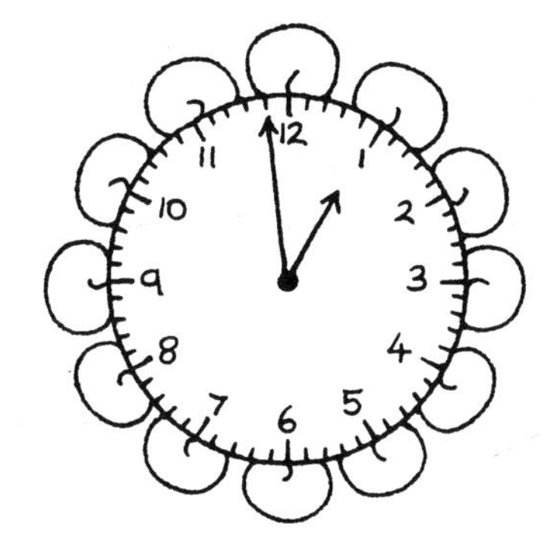
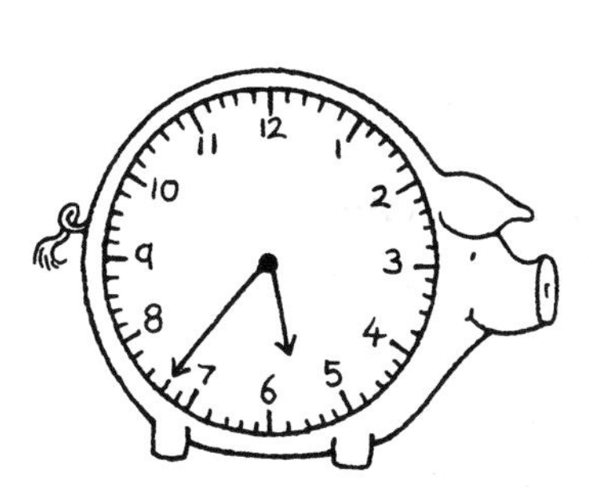

Published 2001 by A & C Black (Publishers) Limited
37 Soho Square, London W1D 3QZ

ISBN 0-7136-5879-7

Editors: Lynne Williamson and Marie Lister

The authors and publishers would like to thank Madeleine Madden and Corinne McCrum for their advice in producing this series of books.

A CIP catalogue record for this book is available from the British Library.

Printed in Great Britain by Caligraving Ltd, Thetford, Norfolk.

Introduction

Developing Numeracy: Measures, Shape and Space is a series of seven photocopiable activity books designed to be used during the daily maths lesson. They focus on the fourth strand of the National Numeracy Strategy *Framework for teaching mathematics*. The activities are intended to be used in the time allocated to pupil activities; they aim to reinforce the knowledge, understanding and skills taught during the main part of the lesson and to provide practice and consolidation of the objectives contained in the framework document.

Year 4 supports the teaching of mathematics by providing a series of activities which develop essential skills in measuring and exploring pattern, shape and space. On the whole the activities are designed for children to work on independently, although this is not always possible and occasionally some children may need support.

Year 4 encourages children to:

- use, read and write the language of measure, shape, space and time;
- estimate, measure and compare length, mass, capacity and time, and to suggest suitable units and equipment for such measurements;
- record readings from scales to a suitable degree of accuracy, and draw and measure lines to the nearest half centimetre;
- measure and calculate the perimeter and area of simple shapes;
- recognise and describe features of 3-D shapes, and visualise 3-D shapes from 2-D drawings;
- classify polygons and describe their features;
- sketch the reflections of simple shapes in a mirror line;
- read and write the vocabulary related to position, direction and movement, and use the eight compass directions;
- know the relationships between turns and angles measured in degrees.

Extension

Many of the activity sheets end with a challenge (**Now try this!**) which reinforces and extends the children's learning, and provides the teacher with the opportunity for assessment. On occasion, you may wish to read out the instructions and explain the activity before the children begin working on it. The children may need to record their answers on a separate piece of paper.

Organisation

Very little equipment is needed, but it will be useful to have available rulers, scissors, coloured pencils, counters, solid shapes, dice, small mirrors, squared paper, small clocks with moveable hands, and interlocking cubes. You will need to provide copies of this year's calendar for page 34.

The children should also have access to measuring equipment to give them practical experience of length, mass and capacity.

To help teachers to select appropriate learning experiences for the children, the activities are grouped into sections within each book. However, the activities are not expected to be used in that order unless otherwise stated. The sheets are intended to support, rather than direct, the teacher's planning.

Some activities can be made easier or more challenging by masking and substituting some of the numbers. You may wish to re-use some pages by copying them onto card and laminating them, or by enlarging them onto A3 paper.

Teachers' notes

Very brief notes are provided at the foot of each page giving ideas and suggestions for maximising the effectiveness of the activity sheets. These can be masked before copying.

Structure of the daily maths lesson

The recommended structure of the daily maths lesson for Key Stage 2 is as follows:

Start to lesson, oral work, mental calculation	5–10 minutes
Main teaching and pupil activities *(the activities in the* ***Developing Numeracy*** *books are designed to be carried out in the time allocated to pupil activities)*	about 40 minutes
Plenary *(whole-class review and consolidation)*	about 10 minutes

Whole-class warm-up activities

The following activities provide some practical ideas which can be used to introduce or reinforce the main teaching part of the lesson.

Measures

Unit thinking

It is useful for children to have a mental image of the size of different units. Ask them to show and describe how big a centimetre, metre, gram or litre is, for example: *A centimetre is the width of my little finger; A metre is the height of my chair/table; A gram is lighter than my pencil sharpener; A litre is about the same as a large carton of orange juice.*

Number line work

On the board, draw a line to represent 1 km and divide it into tenths. Mark 1 km and 1000 m at one end. Count along together in tenths of a kilometre and in hundred metres, for example: *One tenth of a kilometre, two tenths of a kilometre...* and then *One hundred metres, two hundred metres...* Ask the children to give you equivalent pairs, for example: *One tenth of a kilometre is the same as how many metres? What is 400 metres in kilometres?*

Estimating activities

Ask the children to estimate the length, mass or capacity of objects around the classroom, or use household groceries (with the actual measures masked). Encourage them to suggest a range within which the measurement might fall, for example, between 2 and 2·2 metres. Test their estimates by measuring.

Units of time

Discuss the different units of time and the relationships between them. Make a set of cards showing the following times: 1 century, 100 years, 10 years, 1 decade, 1 year, 365 days, 52 weeks, 12 months, 1 week, 7 days, 1 day, 24 hours, 1 hour, 60 minutes, 1 minute, 60 seconds. Hold up a pair of cards and ask the children to say which is longer or to say *Snap* if they show an equivalent time. More cards can be produced to extend the activity, for example, 15 minutes, 30 minutes, 45 minutes and the equivalent fraction cards.

Shape and space

2-D shape cutting

Hold up a sheet of A4 paper and tell the children that you are going to make a straight cut through the piece of paper to make two separate pieces. Ask: *Which two shapes can I make?* and allow time for the children to discuss with a partner. Point out that different pairs of shapes can be made, such as two rectangles; a triangle and a pentagon; a triangle and a hexagon; two quadrilaterals. Repeat for different-shaped pieces of paper, for example, a circle or a regular hexagon.

Twenty questions

Hide a shape in a bag and ask the children to find out which shape it is by asking questions. You can only answer their questions *yes* or *no*. Challenge them to guess the shape in twenty questions. If working with flat shapes, children could be encouraged to draw the shape on the board before it is revealed.

Symmetry game

On the board, draw a 6 x 6 grid with a vertical, horizontal or diagonal mirror line through the centre. Divide the class into two teams. Give a child from each team a counter of the same colour. The child from Team 1 places the counter on one side of the mirror line, and the child from Team 2 places theirs as a reflection of this counter. Choose two more children and give them each a counter in a new colour. This time, Team 2 should place their counter first. Award points for correct positioning.

Position, direction and co-ordinates

Draw a 5 x 5 co-ordinate grid on the board. Label the lines with numbers so that co-ordinates can be given, for example (4, 3). Draw a copy of the grid on paper. Choose one of the capital letters H, T, I or F and secretly draw it on your duplicate grid. Divide the class into two teams. In turn, a child from each team should choose a co-ordinate. This should be marked with a circle if it is one of the points your letter crosses, or with a cross if it is not. The aim is to be the first team to guess the letter and to write down a list of co-ordinates the letter passes through.

Turnabout

Ask the children to stand up and close their eyes. Call out an instruction, such as: *Turn 90 degrees clockwise* or *Turn 180 degrees anti-clockwise.* Encourage the children to turn slowly and not to look at others.

Balloon pairs

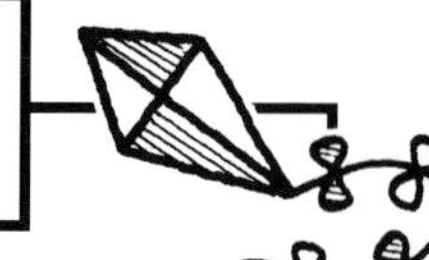

Each balloon must land on the matching measure on the island.

• Write the correct measure on the basket.

m = metres
cm = centimetres

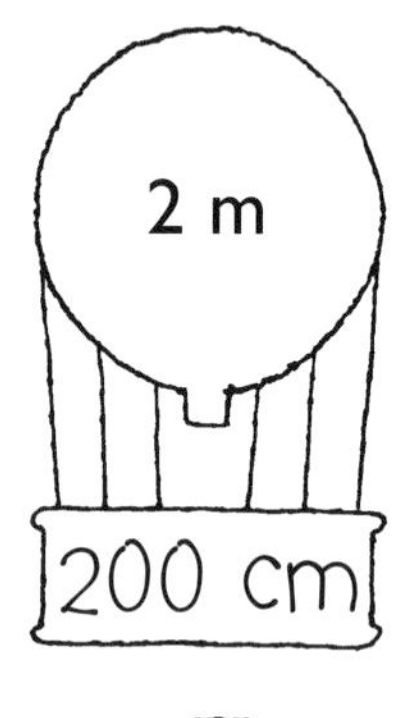

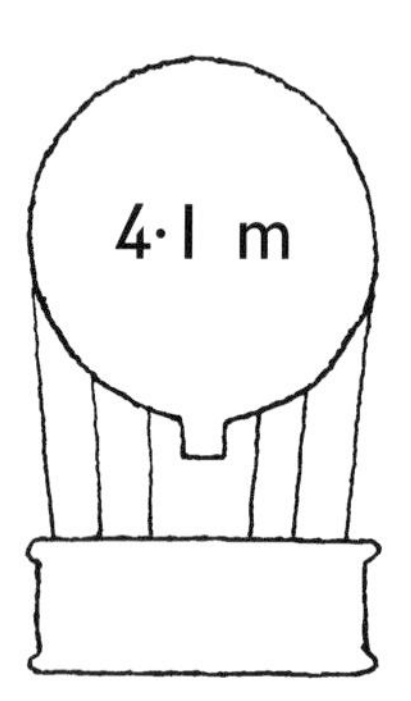

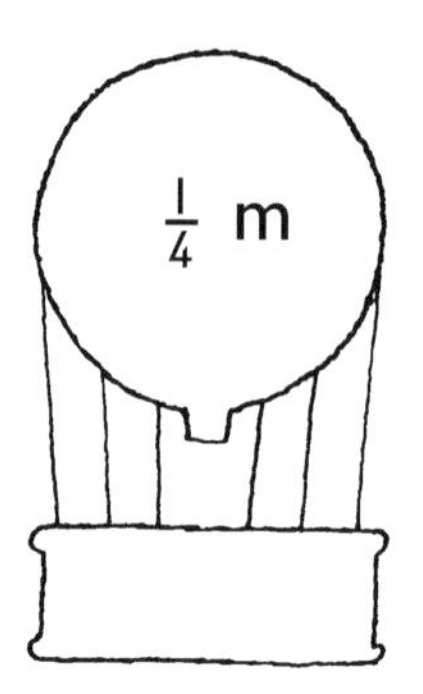

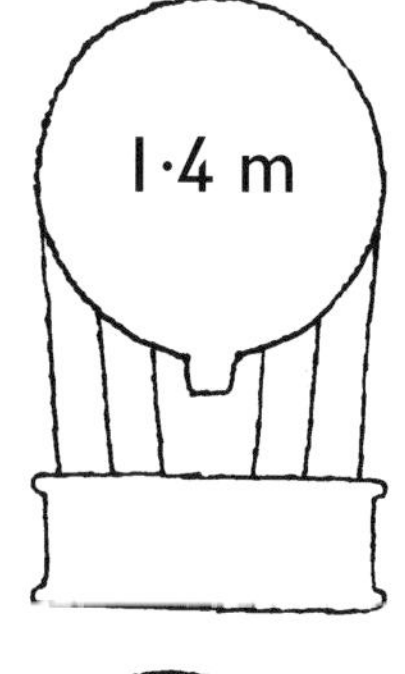

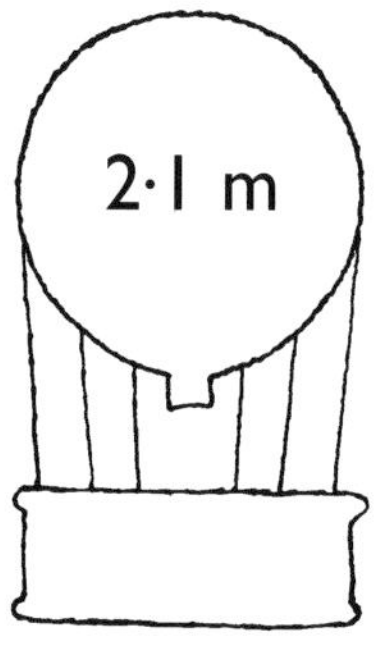

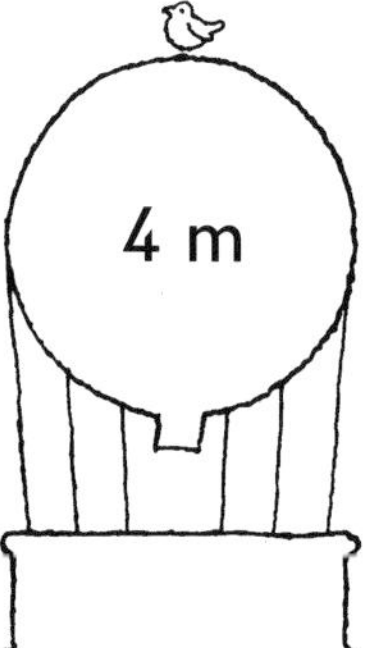

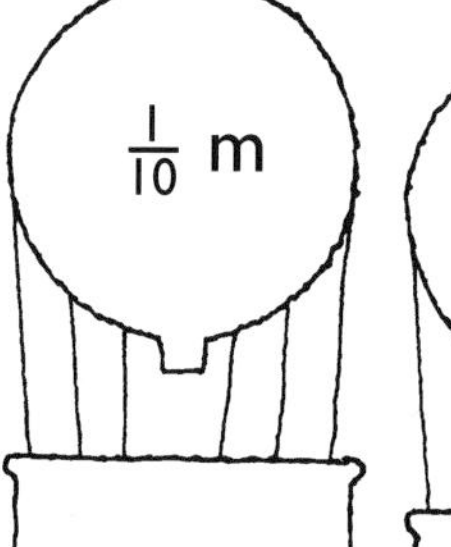

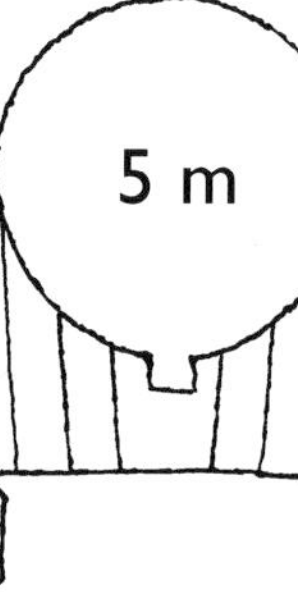

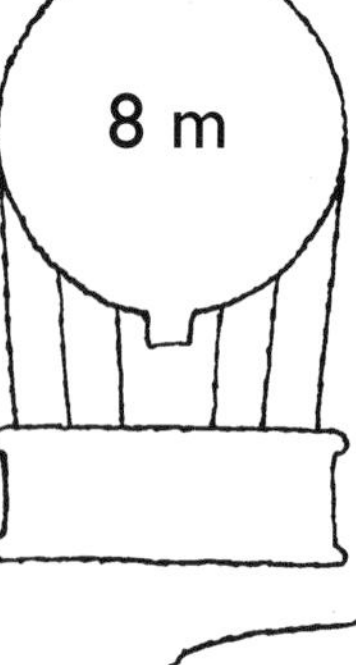

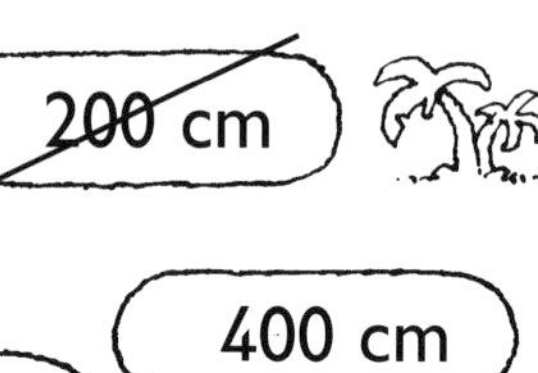

~~200 cm~~ 500 cm 120 cm 400 cm 340 cm 75 cm 140 cm 800 cm 210 cm 10 cm 50 cm 410 cm 25 cm

• Order the lengths in metres. Start with the smallest.

$\frac{1}{10}$ m, $\frac{1}{4}$ m,

Teachers' note Remind the children that there are 100 cm in a metre. Give the children practical experience of using decimal measures in metres through activities such as measuring objects using decimetres and counting along a metre stick in tenths of a metre.

On the beach

• **Write the missing numbers.**

1. 1·6 m = 160 cm

2. 4·4 m = ____ cm

3. 5·2 m = ____ cm

4. 7·1 m = ____ cm

5. 3·3 m = ____ cm

6. 1·9 m = ____ cm

7. 6·1 m = ____ cm

8. 8·2 m = ____ cm

9. 3·2 m = 320 cm

10. ____ m = 280 cm

11. ____ m = 400 cm

12. ____ m = 370 cm

• **Write four more statements of your own.**

Use centimetres **and** metres.

Teachers' note Discuss the sizes of these measurements and encourage the children to estimate or suggest objects of these lengths or heights, for example, '1·6 m is the height of our teacher.'

Nesting birds

- **Cut out the cards. Match the birds to the nests.**
  Some nests may be empty, others may have more than one bird.

km = kilometre
1000 m = 1 km

1 km	0·5 km	$\frac{1}{2}$ km	$\frac{1}{4}$ km
$\frac{1}{10}$ km	$1\frac{1}{2}$ km	6 km	3 km
2 km	$\frac{3}{4}$ km	4 km	10 km
0·25 km	1·5 km	0·1 km	0·75 km
100 m	3000 m	250 m	750 m
10 000 m	2000 m	4000 m	500 m
1000 m	1500 m	10 m	6000 m

Teachers' note Remind the children that there are 1000 m in 1 km and discuss the abbreviations 'km' and 'm'. As an extension, ask the children to order the lengths in kilometres, starting with the longest.

Master your measures

- **Colour the** unit **you would use to measure each item.**

mm = millimetre
10 mm = 1 cm

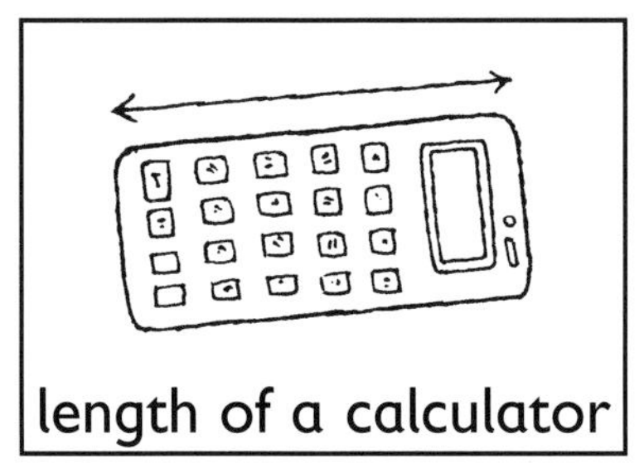

length of a calculator

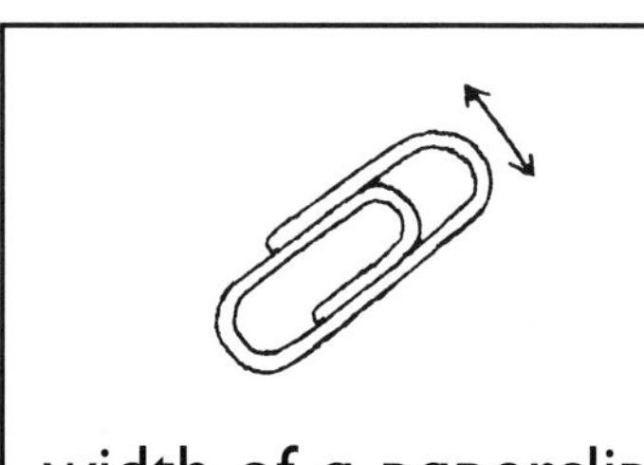

width of a paperclip

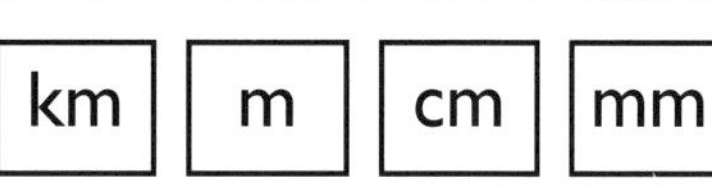

height of a mountain

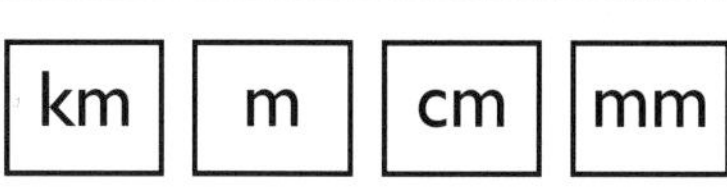

depth of a lake

thickness of a book

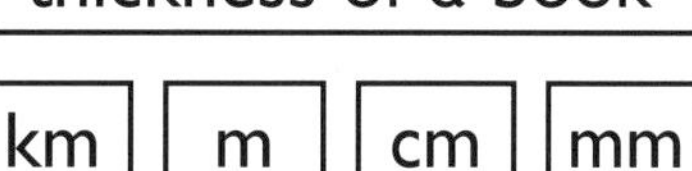

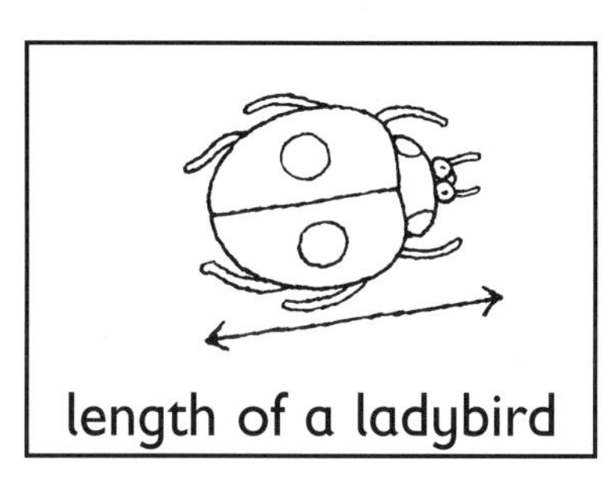

length of a ladybird

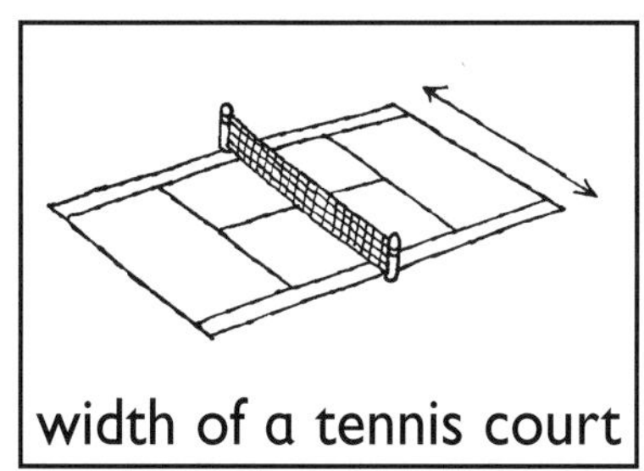

width of a tennis court

km | m | cm | mm

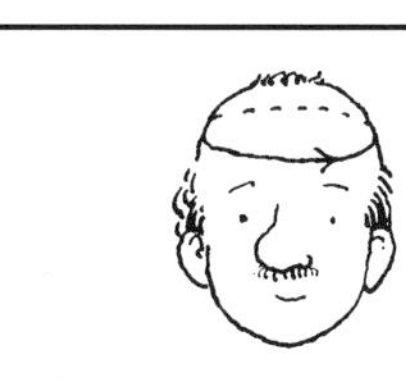

distance around a person's head

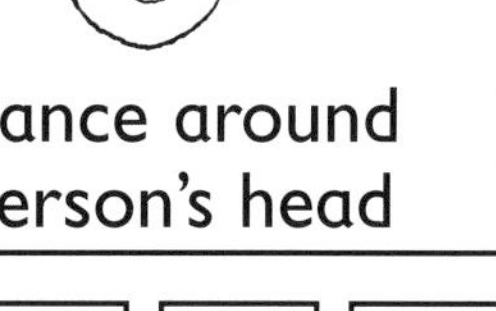

height of a house

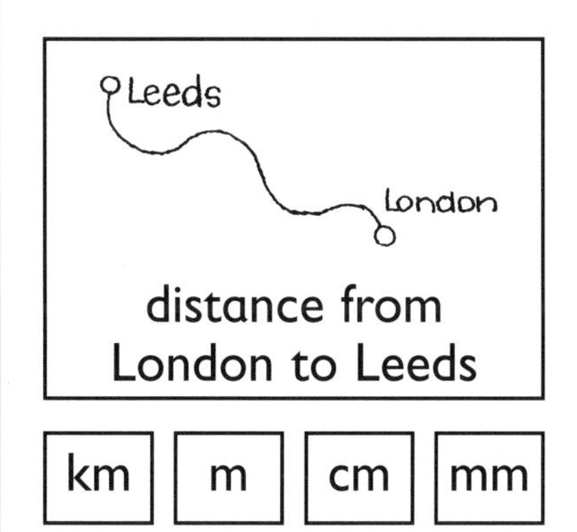

distance from London to Leeds

km | m | cm | mm

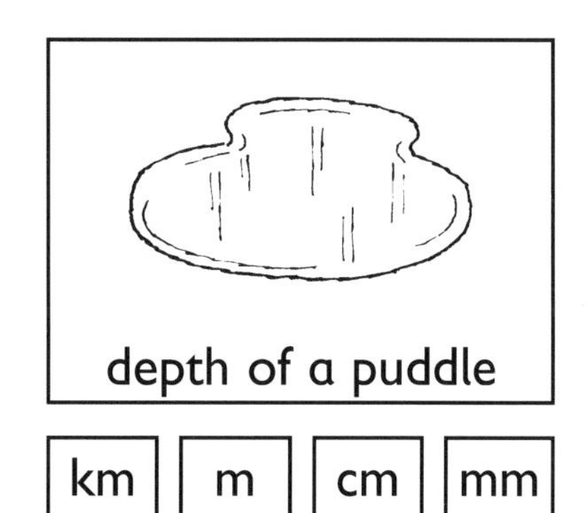

depth of a puddle

km | m | cm | mm

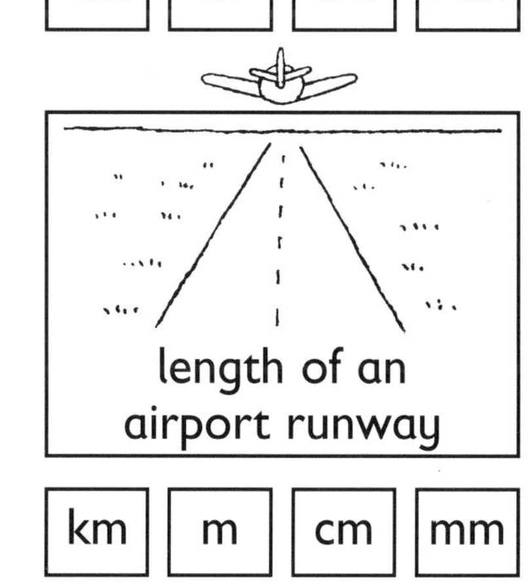

length of an airport runway

km | m | cm | mm

Teachers' note The children could also make and compare estimates for these measurements, for example: 'I think the length of a ladybird could be about...'

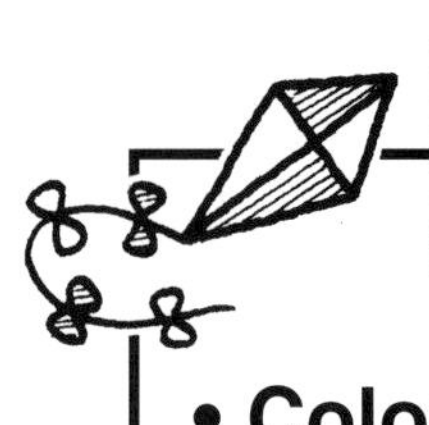

Best estimates

- **Colour the best estimate.**

15 cm / ½ m / 70 cm	length of a computer keyboard
5 cm / 20 cm / 10 cm	height of a mug
5 km / 20 km / 200 km	distance a person can walk in one hour
88 mm / 18 mm / 55 mm	length of a playing card
300 km / 30 km / 3000 km	distance from London to Paris
3 mm / 6 mm / 1·2 cm	thickness of a pound coin
2000 m / 100 m / 20 m	how far a good sprinter runs in 10 seconds
5 m / 1 km / 100 m	width of a football pitch

- **A tennis court has a length of about 20 m. Estimate its perimeter.** ________

Teachers' note The children should be encouraged to estimate the lengths, heights, and so on of a range of large and small items. Best estimates should be identified. Before beginning the extension activity, ensure that the children understand 'perimeter' and how to work it out. If appropriate, the children could work in pairs to discuss their ideas.

Spike the hedgehog

- **Use a ruler to measure the length of each of Spike's prickles.**
- **Write your answer to the nearest** half a centimetre**.**

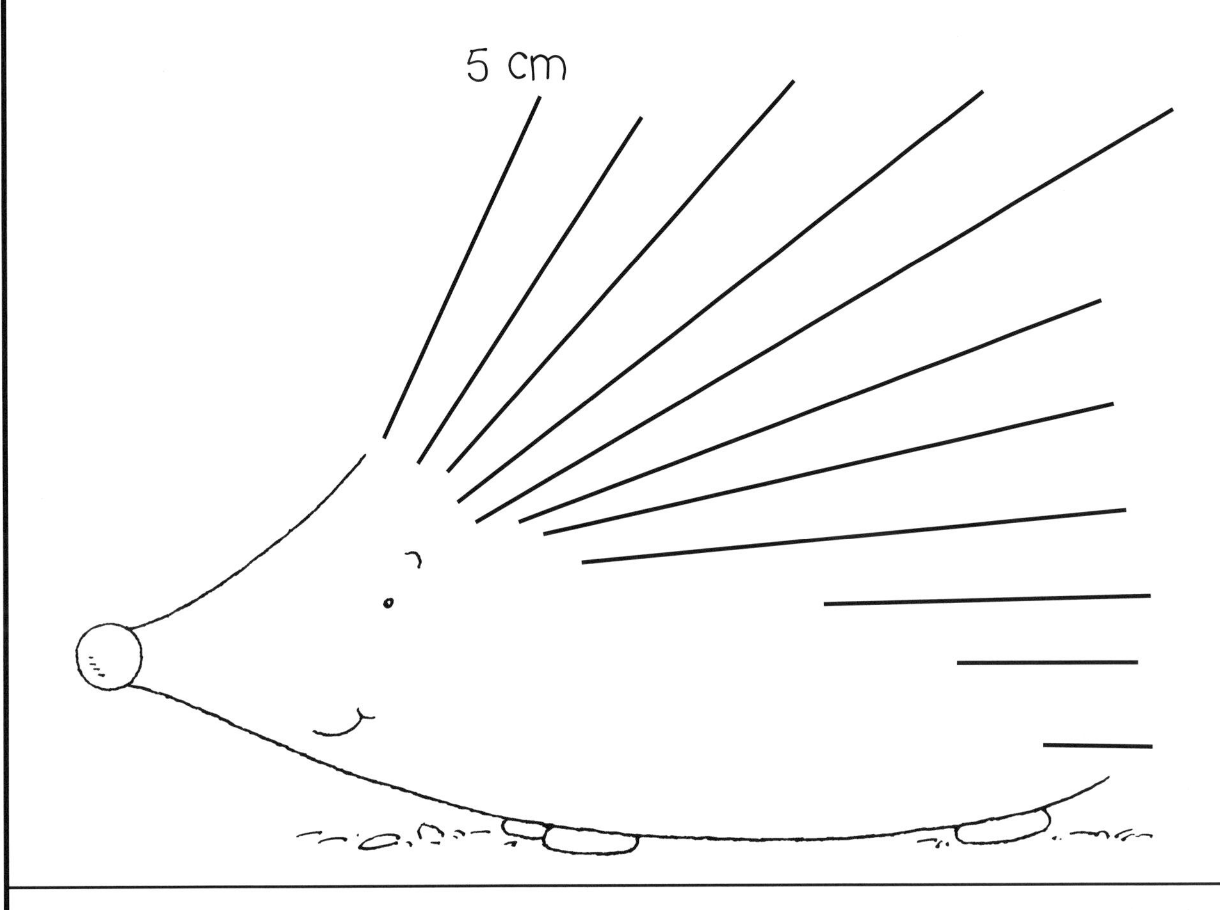

- **Draw these lengths on Spike's friend.**

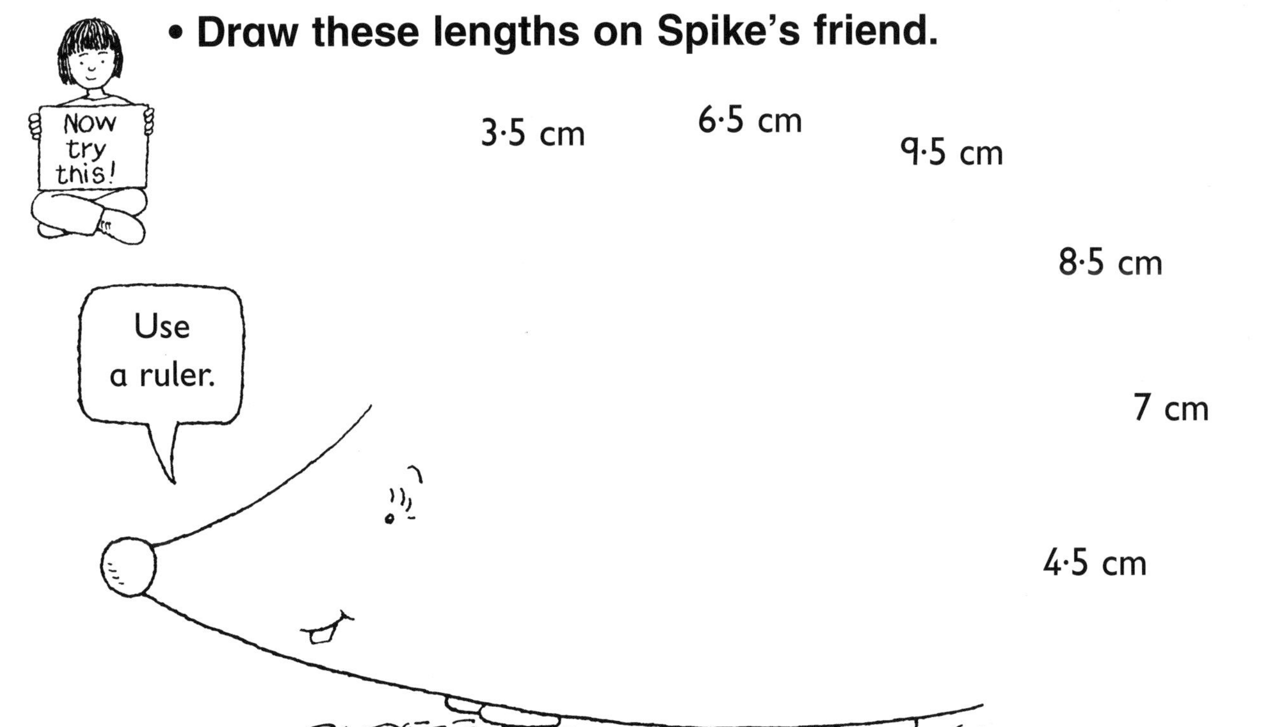

Teachers' note Tell the children to write the unit 'cm' when recording lengths and remind them that the notation can be, for example, $3\frac{1}{2}$ cm or 3·5 cm. Ensure that the children correctly position the ruler on the lines.

Stripy T-shirt

- **Use a ruler to measure each stripe on the T-shirt.**
- **Write your answer on the stripe to the nearest** half a centimetre**.**

- **Draw two more stripes on the T-shirt measuring** 6·5 cm **and** 7·5 cm**.**

- **Now measure each line around the edge of the T-shirt.**
- **What is the total** perimeter**?** ______

Teachers' note Remind the children that half a centimetre can be written as $\frac{1}{2}$ cm or 0·5 cm. For the extension activity, ensure that the children understand 'perimeter' and how to work it out.

Insect tug-of-war

The ants and the beetle are having a tug-of-war.

- Read each scale. How far from the ant hill is the spot on the leaf?

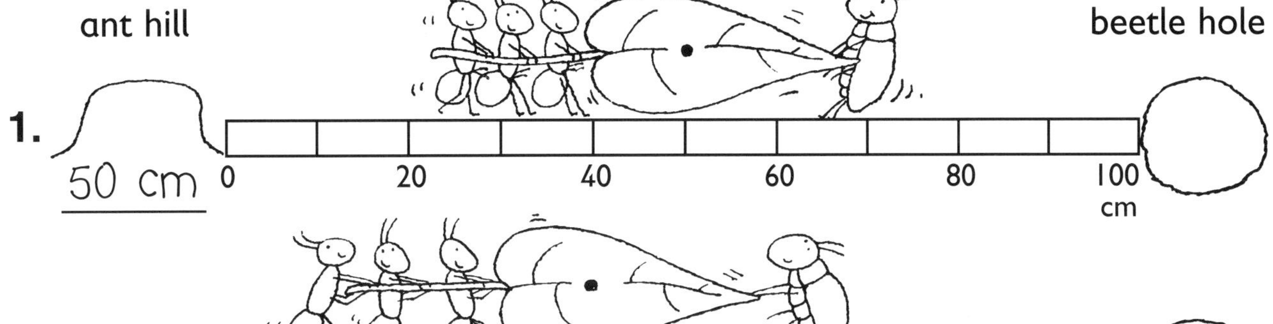

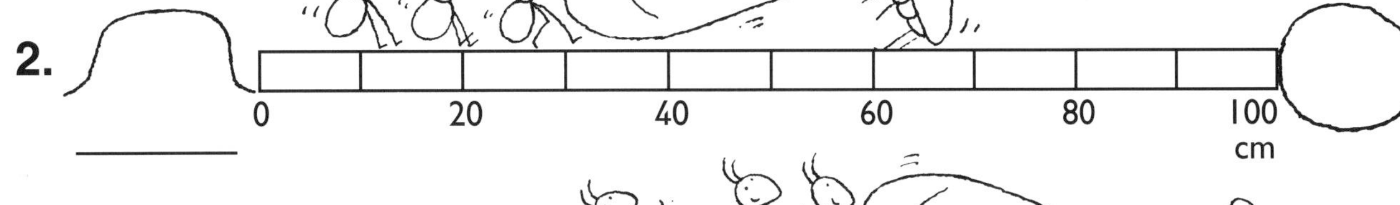

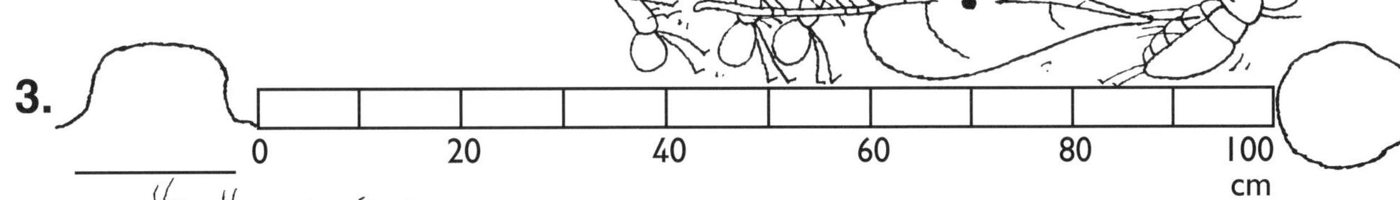

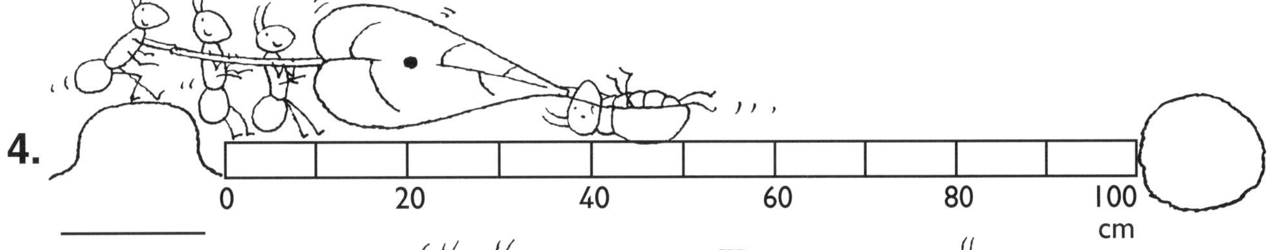

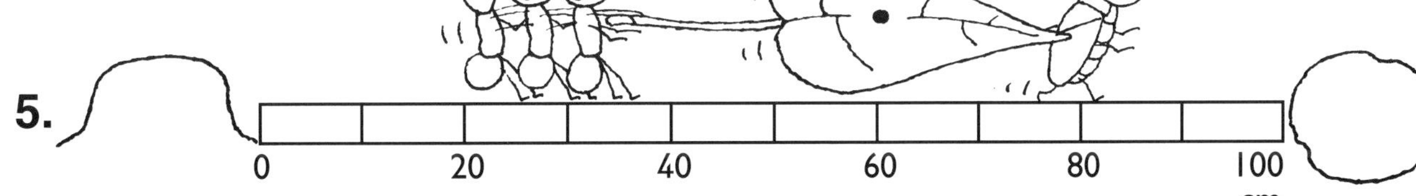

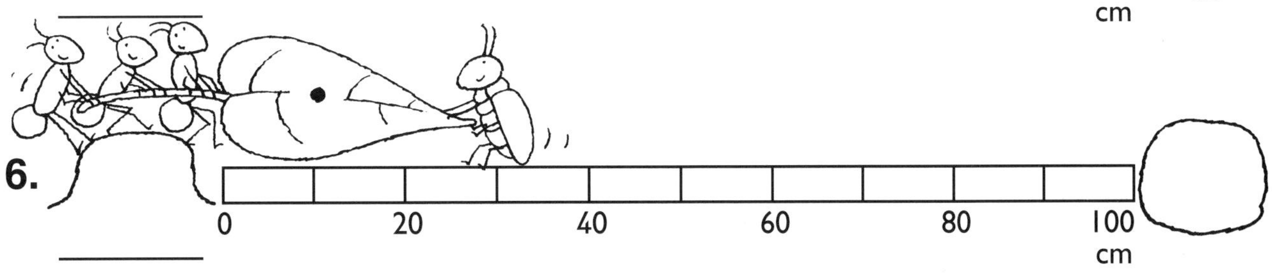

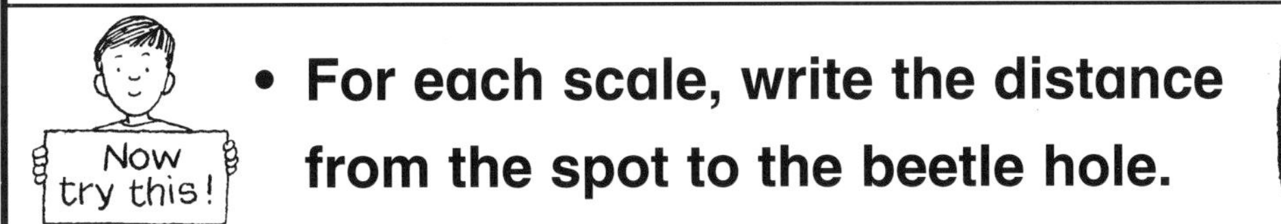

- **For each scale, write the distance from the spot to the beetle hole.**

Teachers' note Remind the children to write the unit 'cm' for each answer. The numbers on the scales can be masked and altered to provide a flexible resource.

Tug-of-war game

• Play this game with a partner. Choose an end and write your name.

☆ Cut out the cards. Put the arrow on 25 cm. Put the other cards face down.
☆ Take turns to pick a card. Move that many places towards your end.
☆ Write down the number you land on after each turn.
☆ The winner is the player who reaches his or her end first.

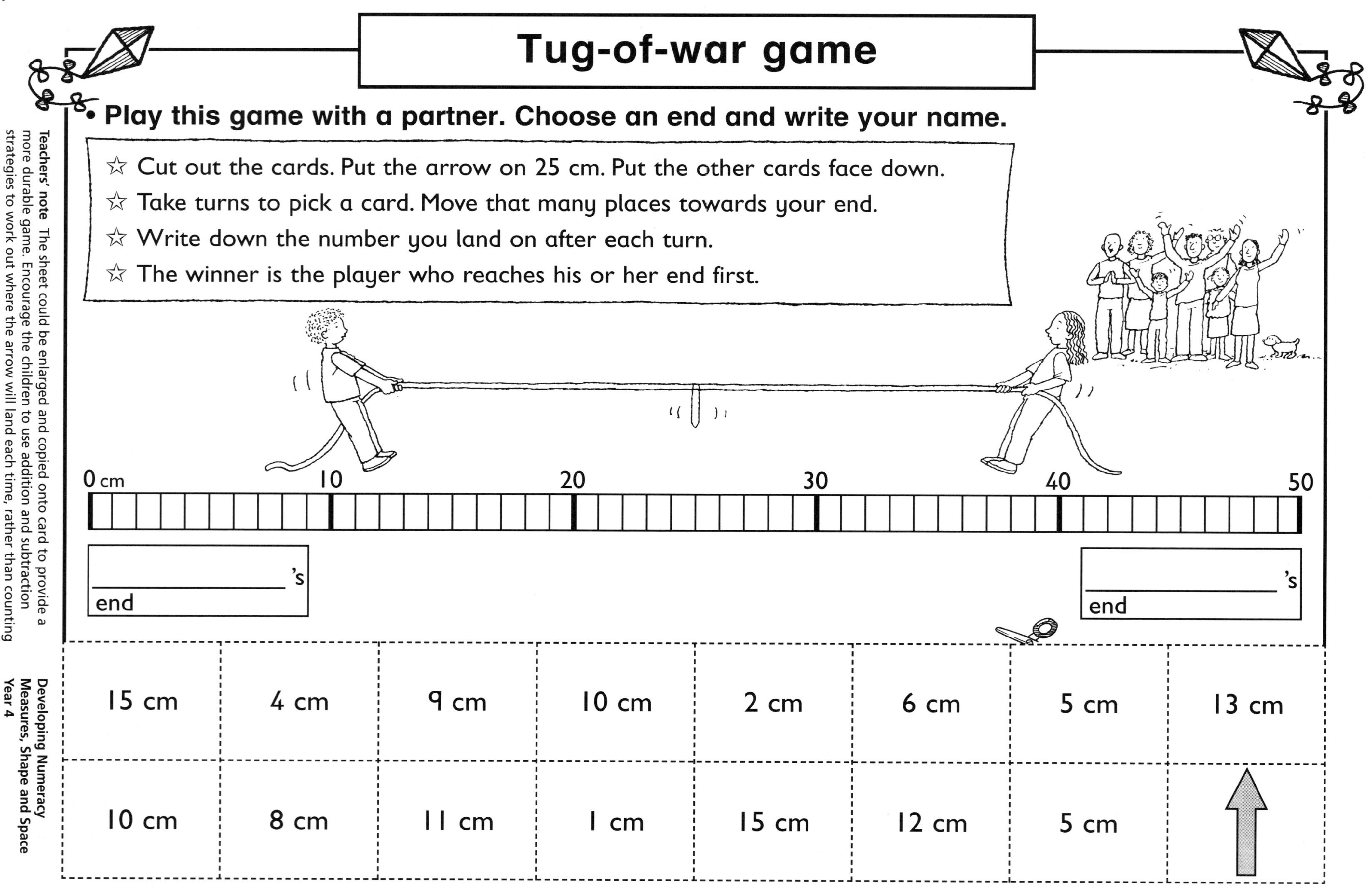

15 cm	4 cm	9 cm	10 cm	2 cm	6 cm	5 cm	13 cm
10 cm	8 cm	11 cm	1 cm	15 cm	12 cm	5 cm	↑

Teachers' note The sheet could be enlarged and copied onto card to provide a more durable game. Encourage the children to use addition and subtraction strategies to work out where the arrow will land each time, rather than counting along the scale in ones.

How heavy?

- You can use g or kg to measure how heavy these items are. Write which is best.

g = gram
kg = kilogram

1. kitchen roll — g

2.

3.

4.

5.

6.

7.

8.

9.

10.

- Estimate how heavy each item is.
- Compare your estimates with a partner.

Teachers' note It would be useful to have some weights and a range of items similar to those above to enable the children to gain an appreciation of the mass of the items.

Dumb-bells

• **Fill in the missing numbers so that each end of the dumb-bell shows the same** mass.

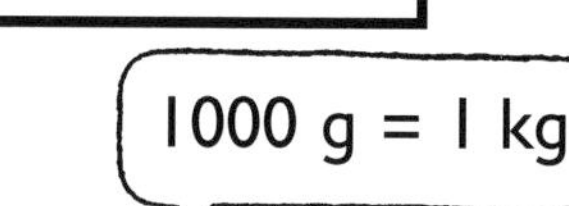

1.

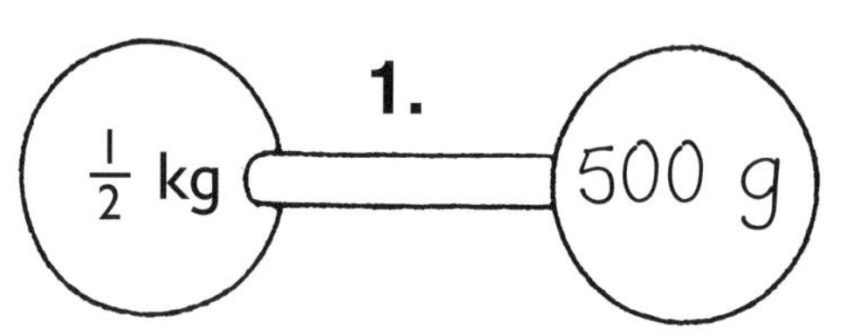

2.

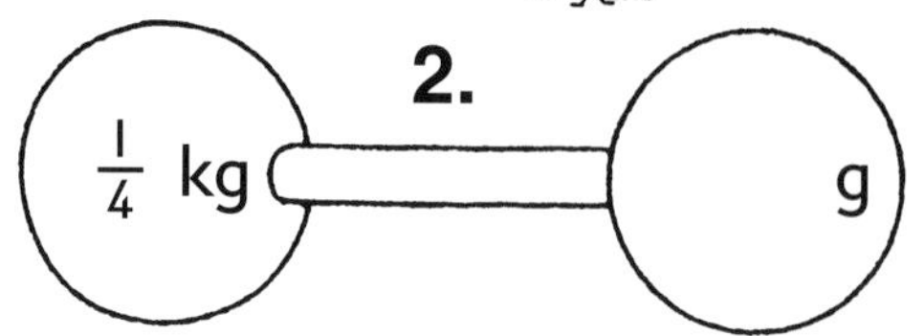

3.

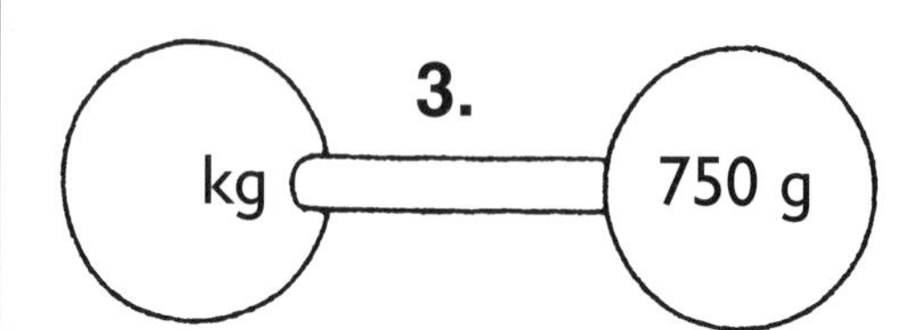

4.

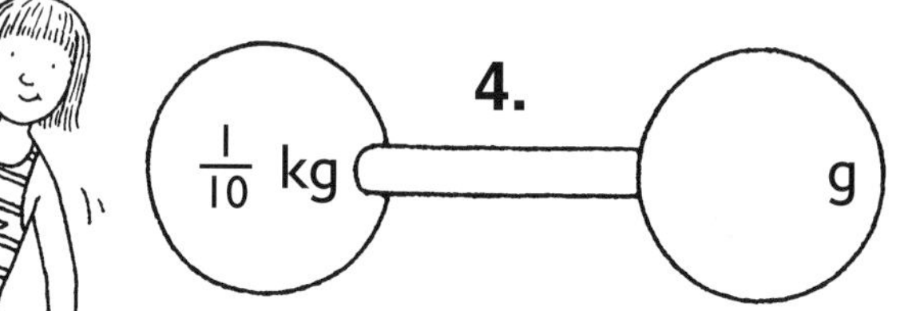

5.

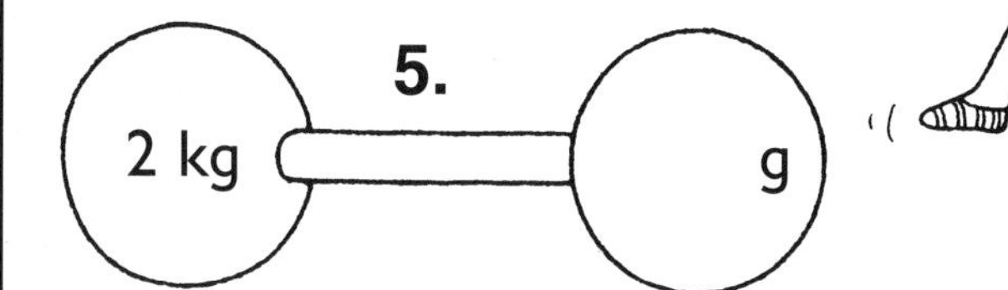

6.

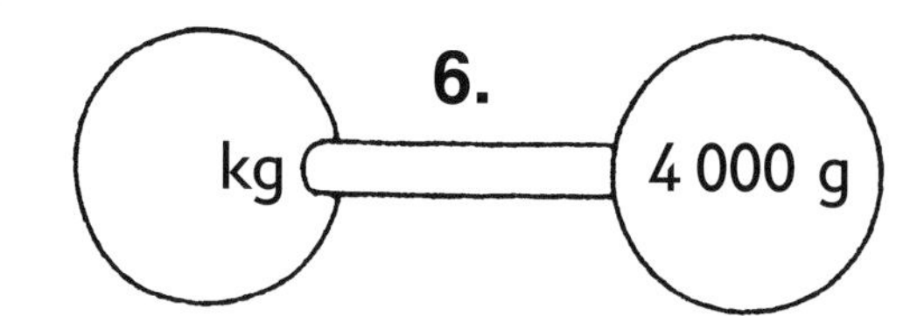

7.

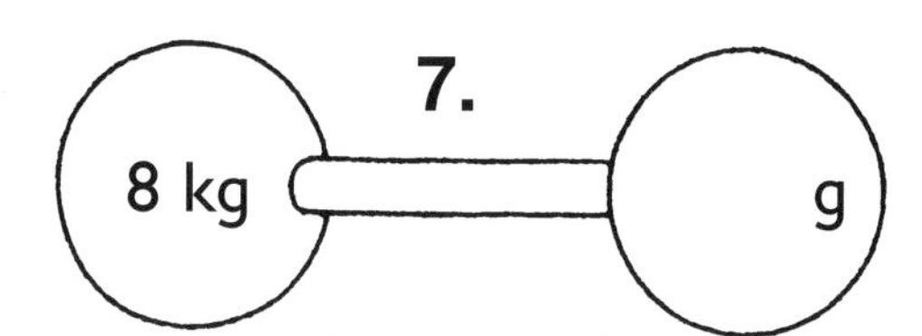

8.

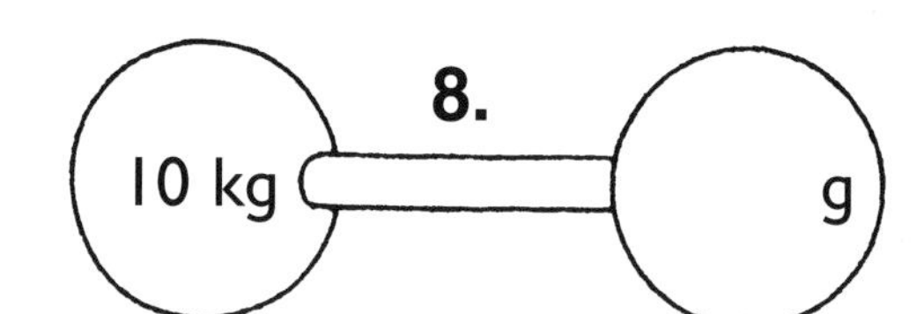

9.

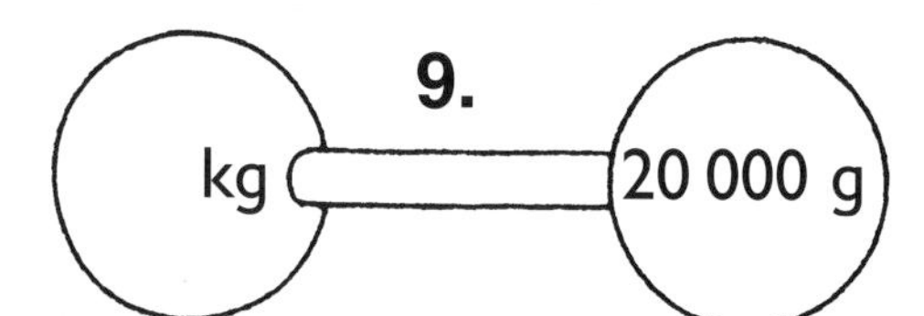

10.

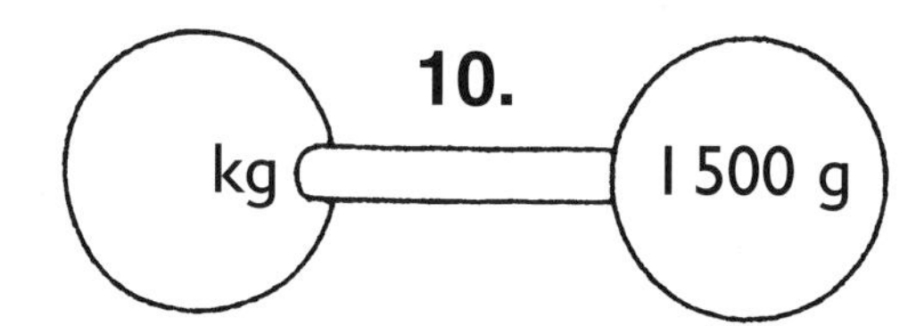

11.

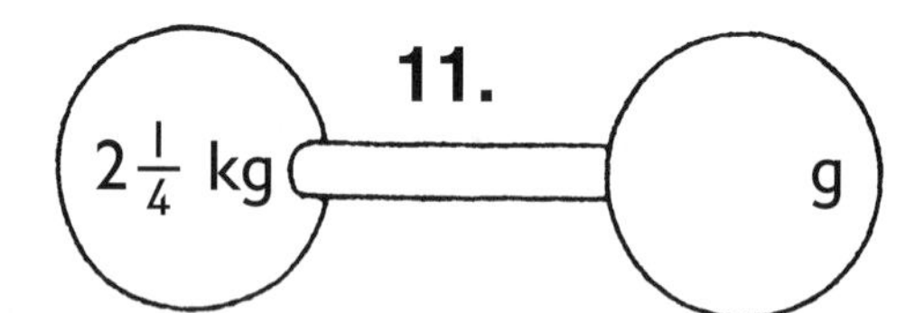

12.

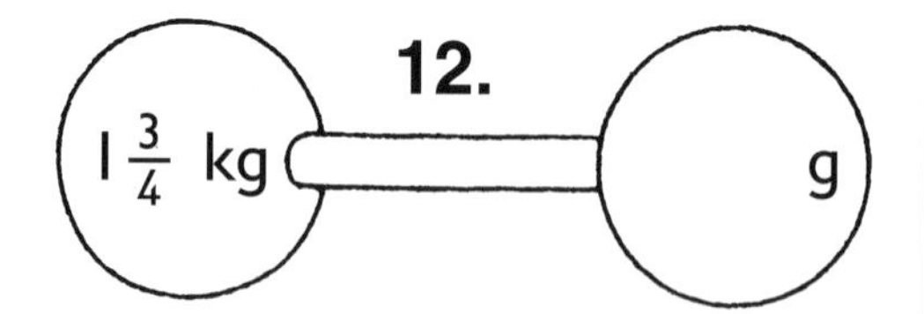

• **Which dumb-bell shows:**

the heaviest mass? _____

the lightest mass? _____

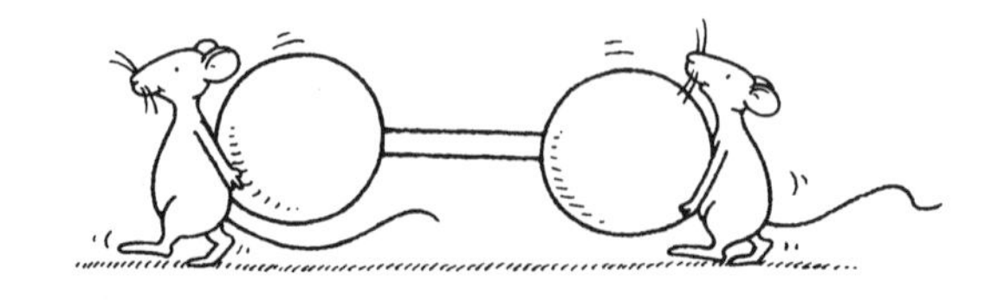

Teachers' note You may wish to mask the last three questions for some children. All the values can be masked and altered to provide an extension challenge or a simplification. Encourage the children to use the mathematical word 'mass' for these values. As a further extension, the children can place the masses in order.

Masses of vegetables

• **Read the scales. Write the mass of the vegetables.**

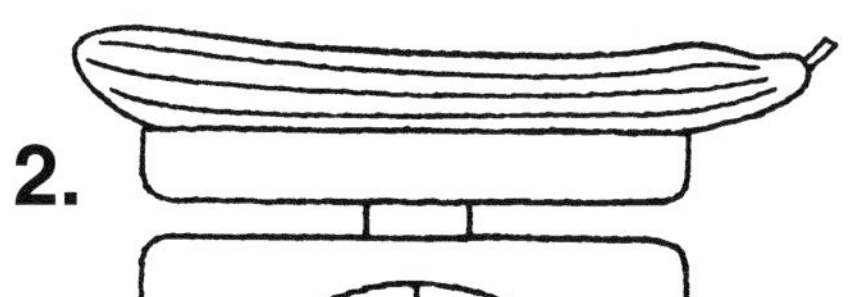

1.

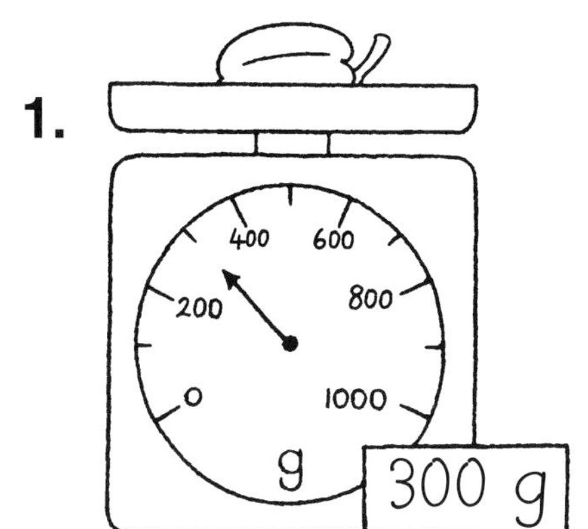

2.

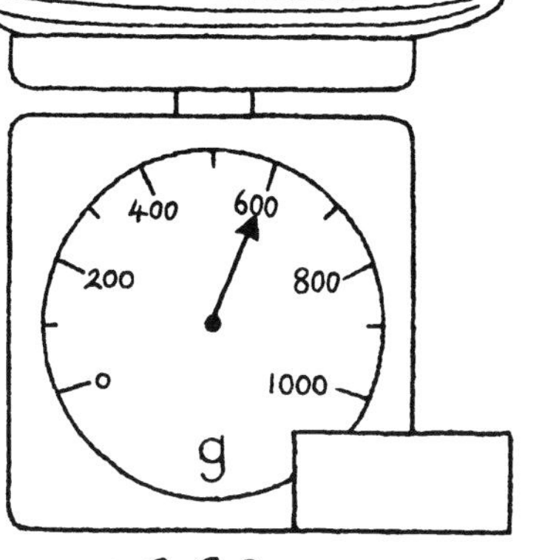

3.

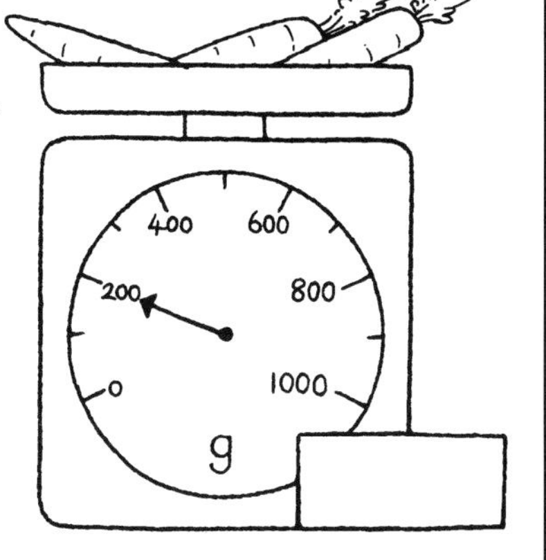

4.

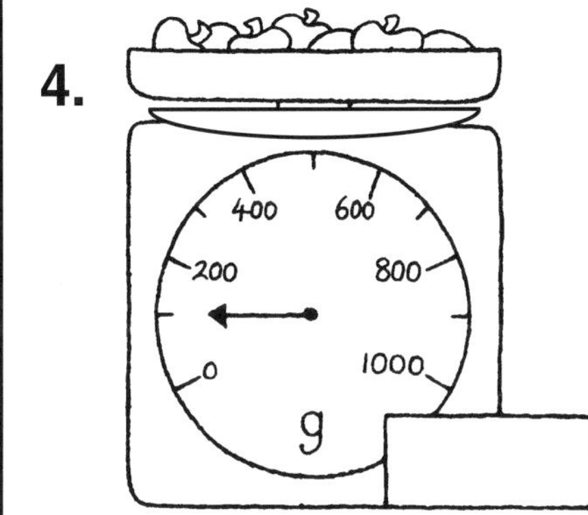

5.

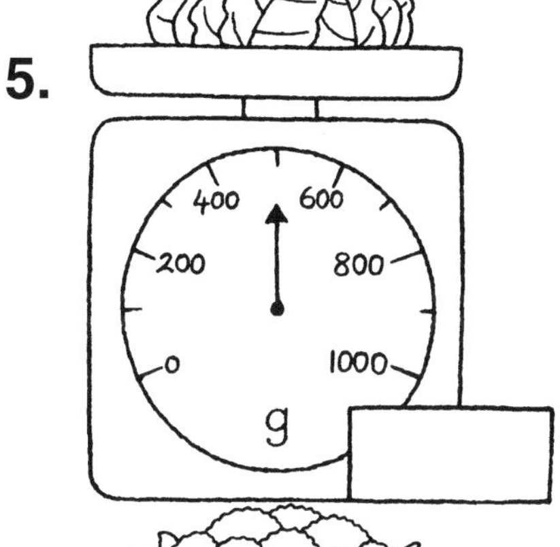

6.

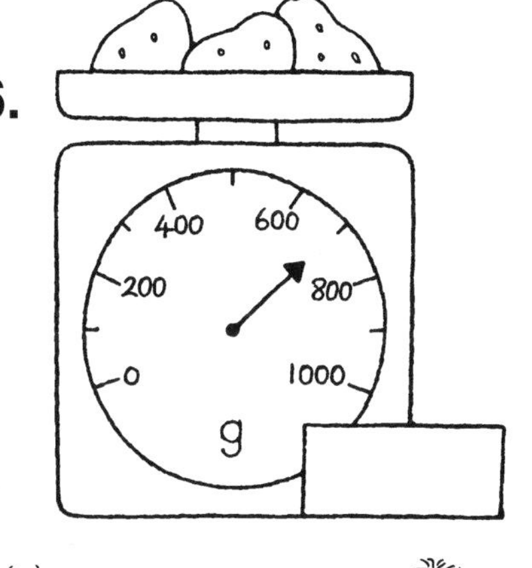

7.

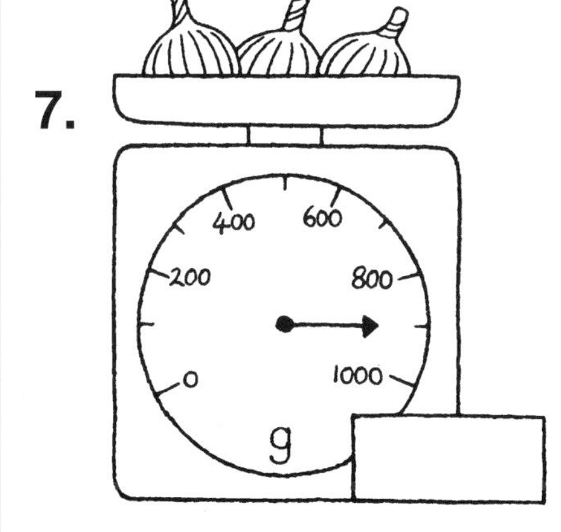

8.

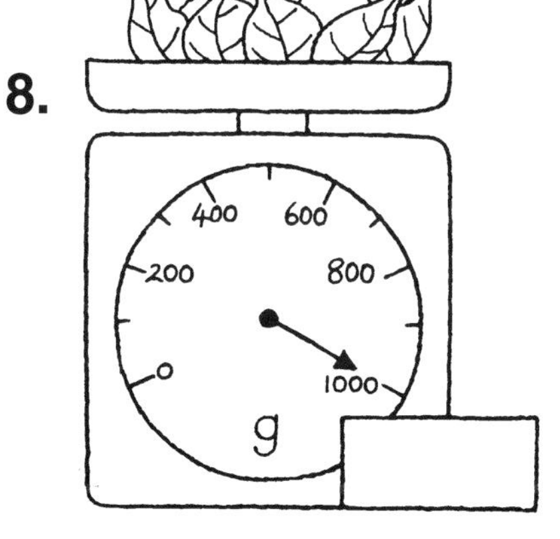

9.

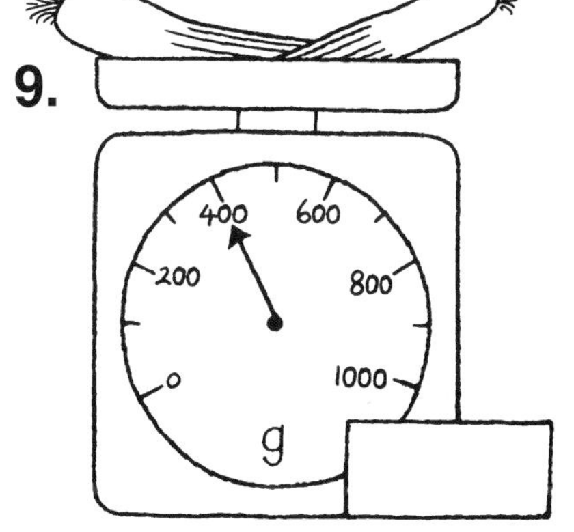

• **Draw arrows on the scales to show these masses.**

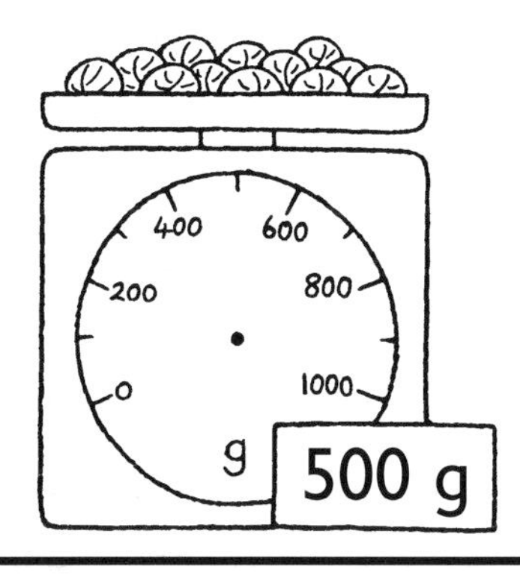

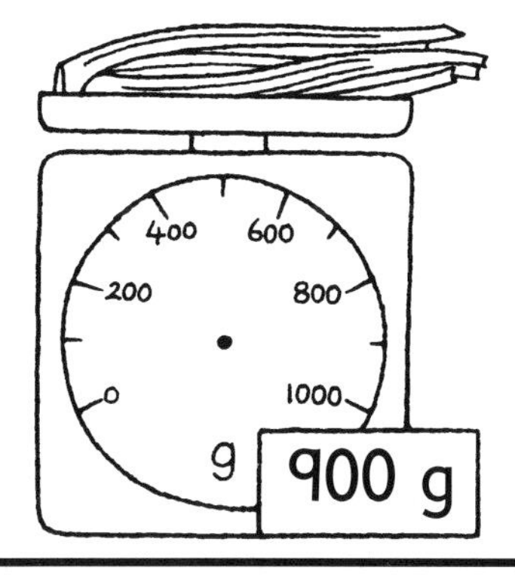

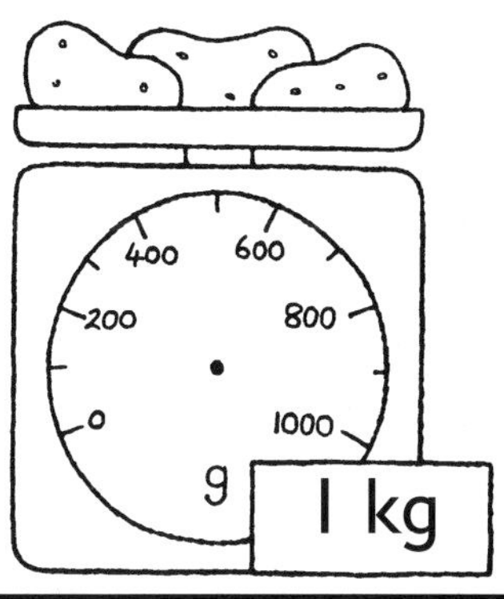

Teachers' note The arrows can be masked before photocopying to provide a more flexible resource. Remind the children to write the unit 'g' together with the number, and to look at the numbers on the scale to work out how much each interval is worth.

Water pistols

- **Join each water pistol to the target that shows the same amount.**

l = litre
ml = millilitre
1000 ml = 1 l

5 l

3000 ml

3 l

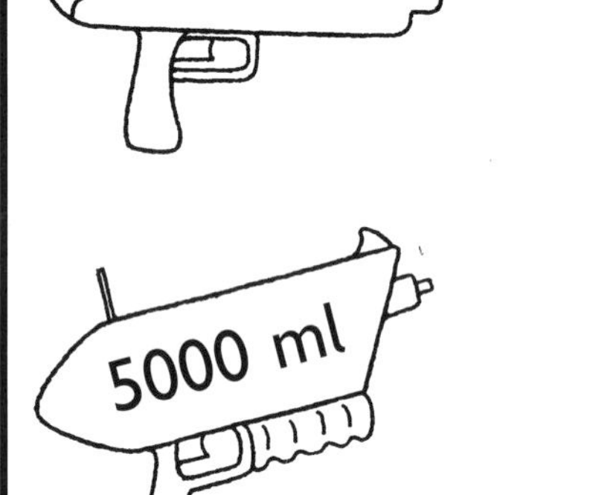

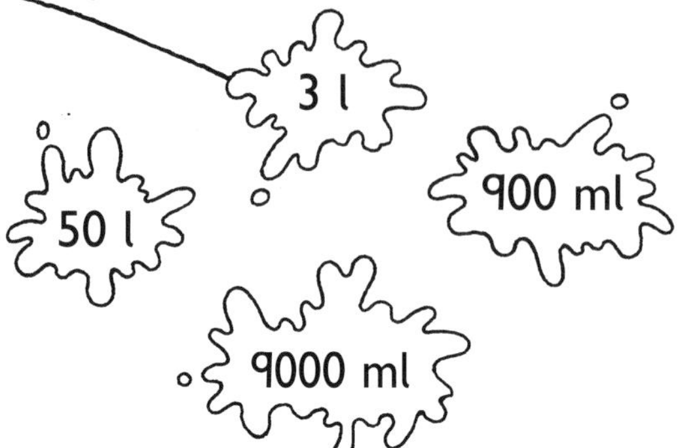

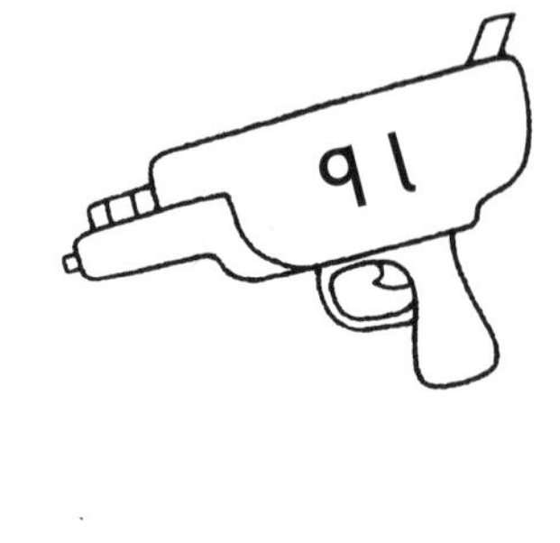

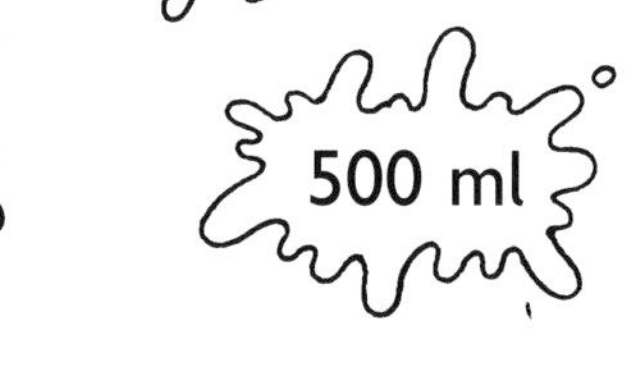

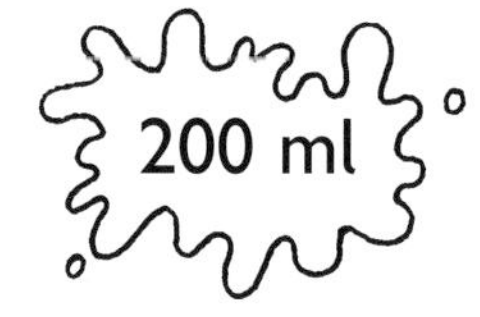

2000 ml

700 ml

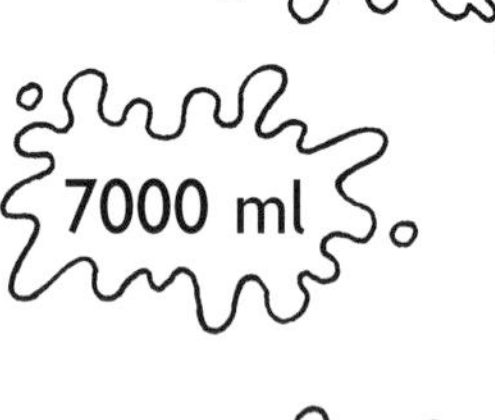

6 l

4000 ml

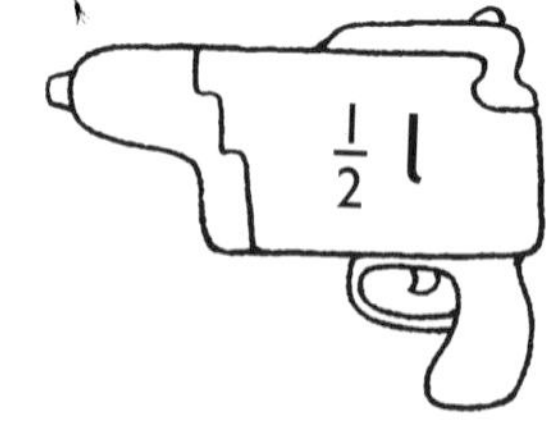

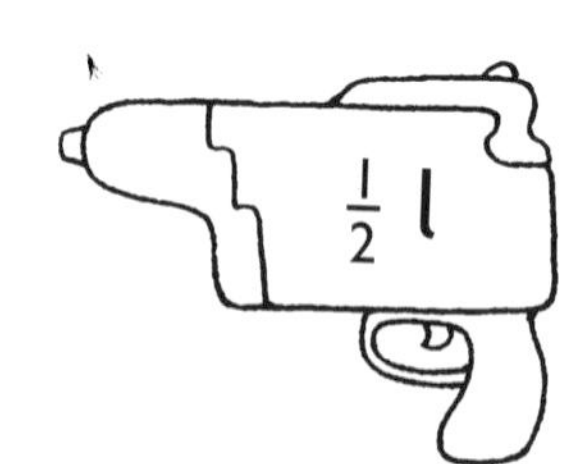

60 l

- **Write the number of** millilitres **in:**

$\frac{1}{8}$ l ________ $\frac{1}{4}$ l ________ $\frac{3}{8}$ l ________ $\frac{3}{4}$ l ________

Teachers' note Before beginning the activity, give the children practical experience of using measuring jugs marked in both 'ml' and 'l', to establish that they understand the relationship between the units. As a further extension, the children could write equivalent amounts for the targets that are not matched.

MA

Jez and his jet thrusters

- **Write how many** millilitres **of fuel in each jet thruster.**

1.

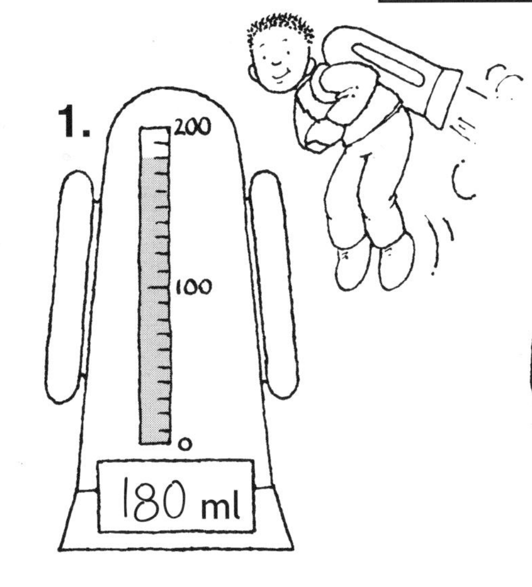

2.

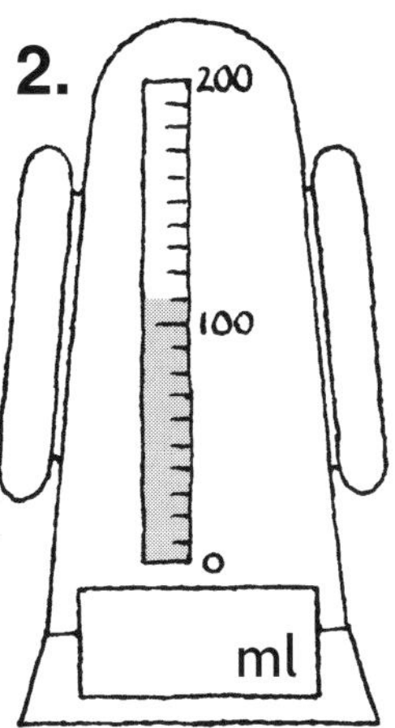

3.

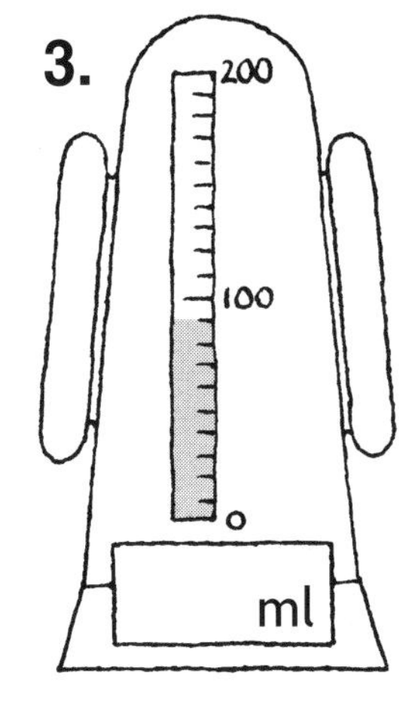

4.

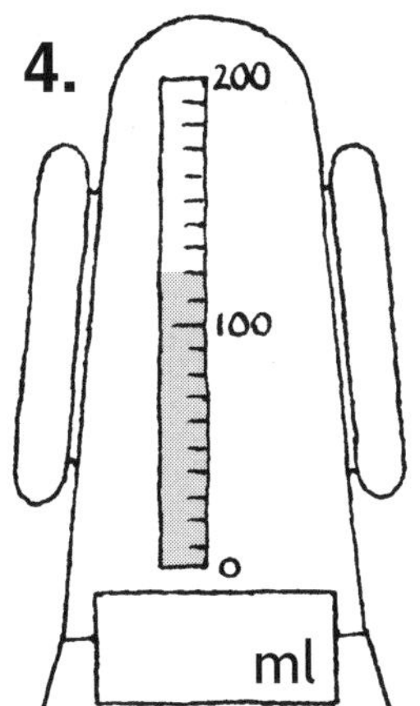

5.

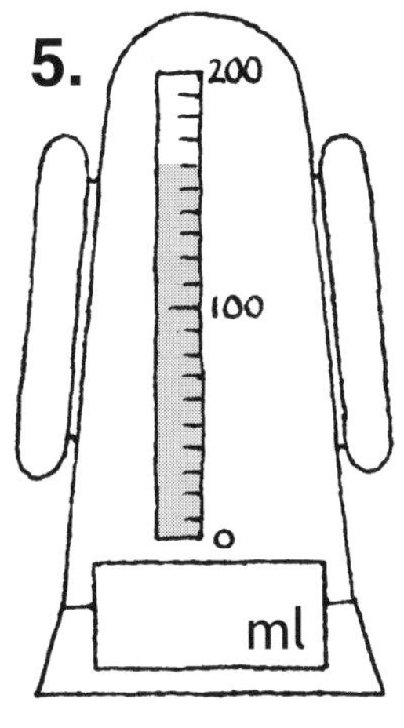

6.

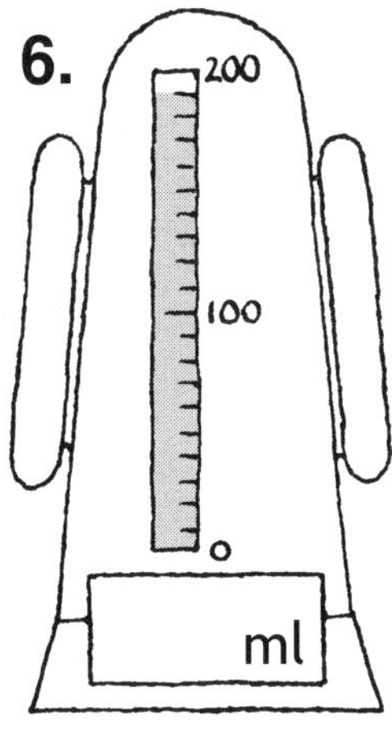

7.

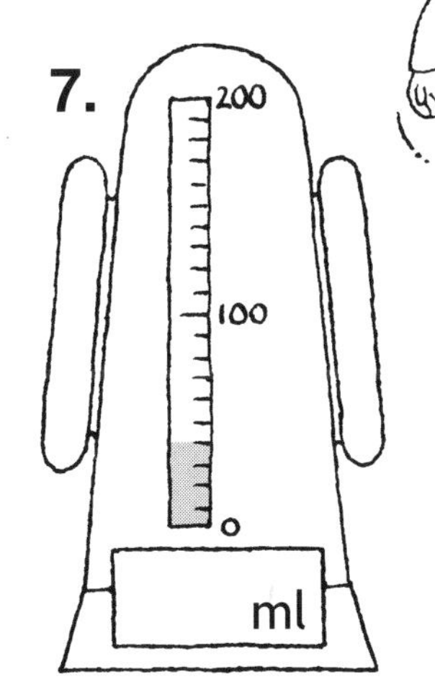

8.

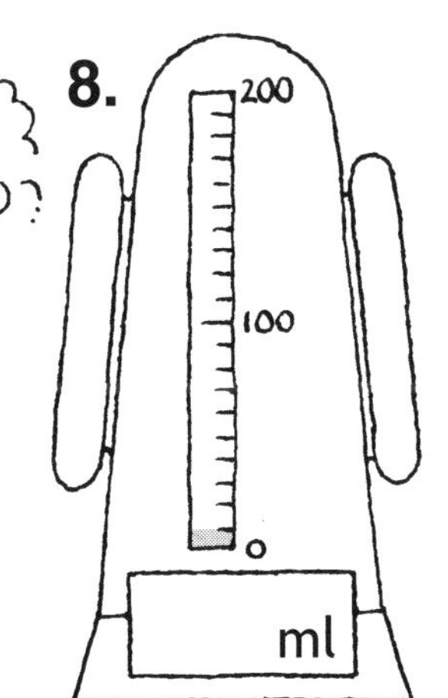

9.

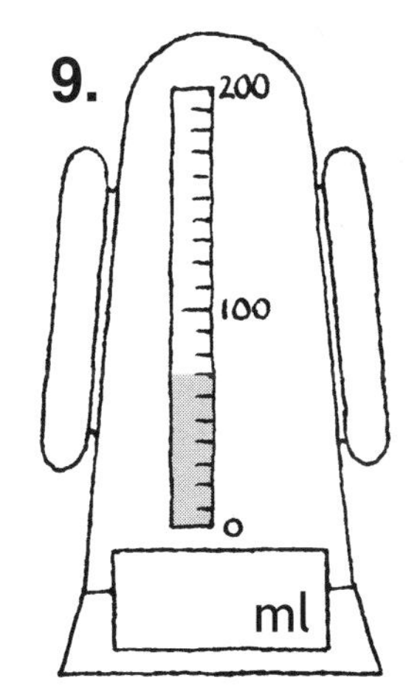

- **Write the jet thrusters in order. Start with the least amount of fuel.**

8, 7, ______________________________

Teachers' note Discuss the amounts, encouraging the children to show and estimate similar amounts of water in containers, for example: 'The fuel in jet 1 is about half the amount of water in this mug.'

Container capacities

How much does each container hold?

• Colour the best estimate **.**

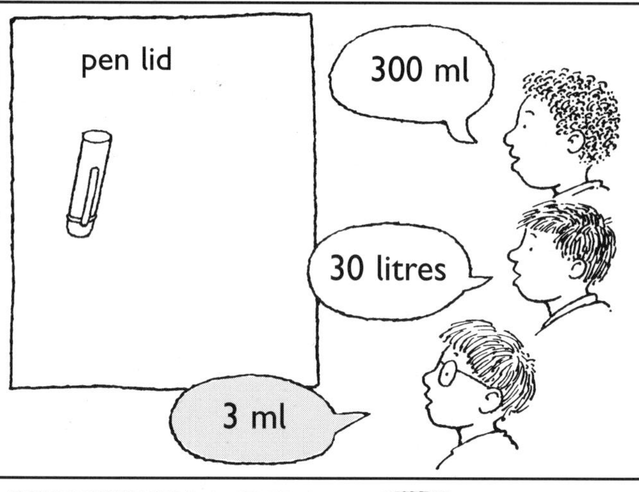

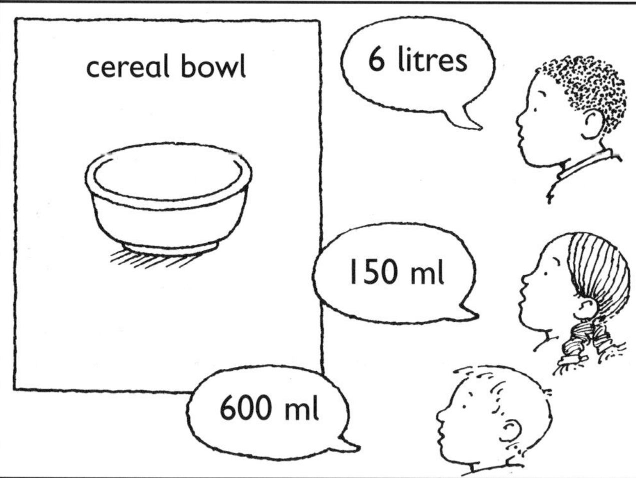

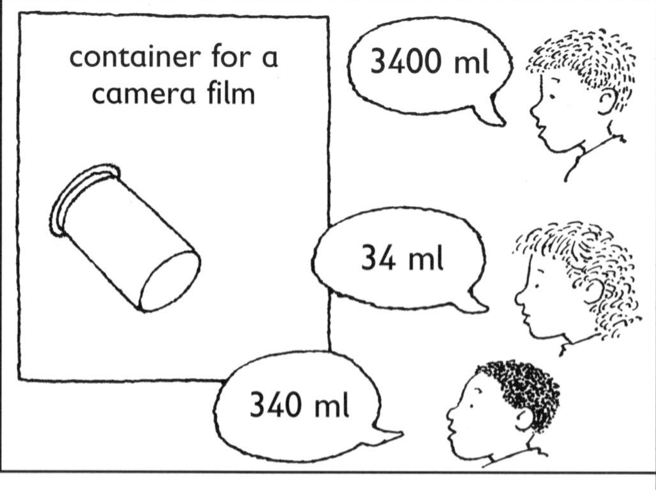

• Draw an object that has a capacity of about:

300 ml   1 litre   10 litres

Teachers' note The children could work in pairs to discuss their ideas. Provide plenty of practical experience of measuring capacities using litres and millilitres. It would be useful to have examples of objects and measuring jugs for the children to refer to and handle.

Control panel

• **Read the dials on the control panel. Write each reading.**

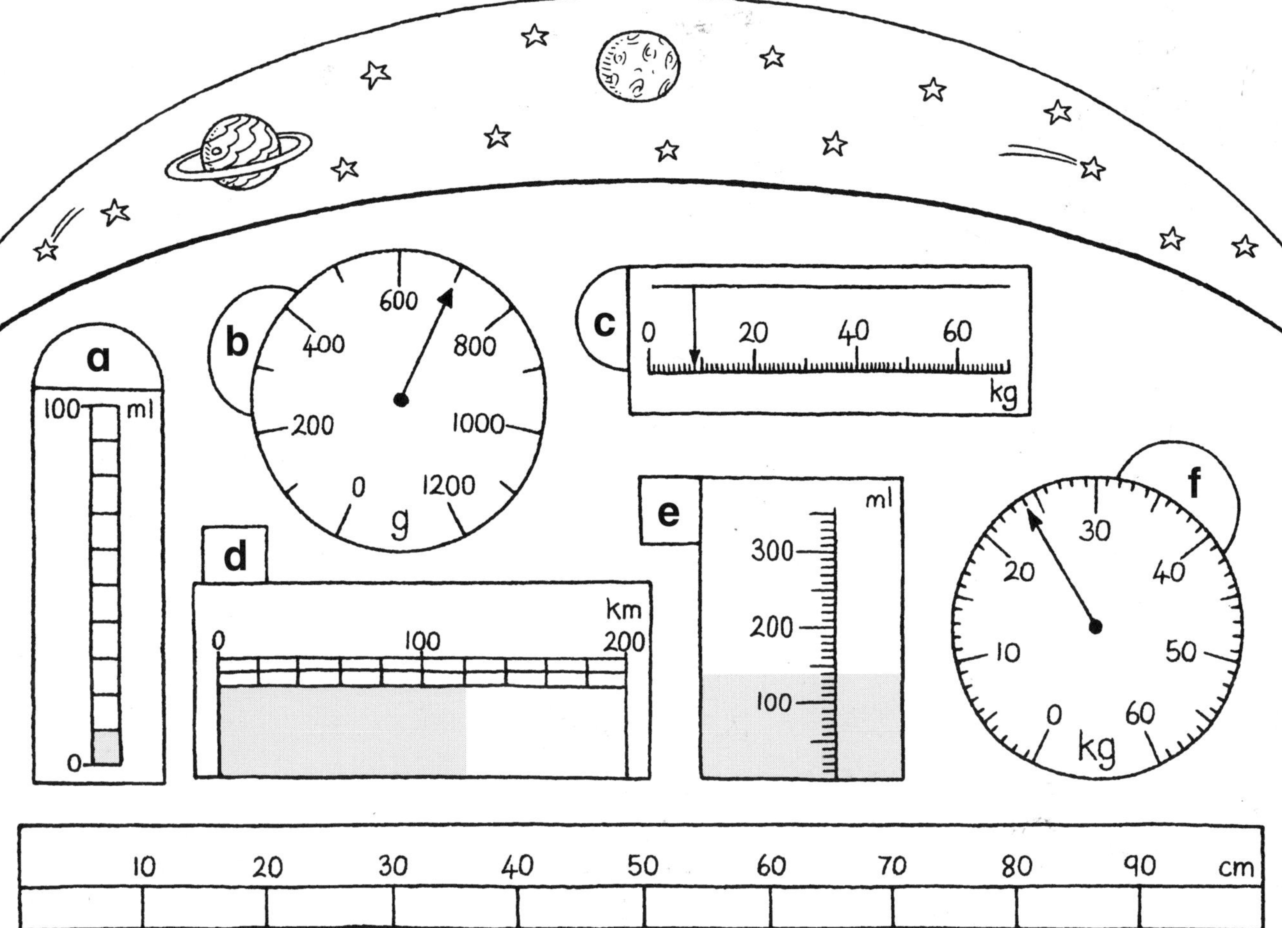

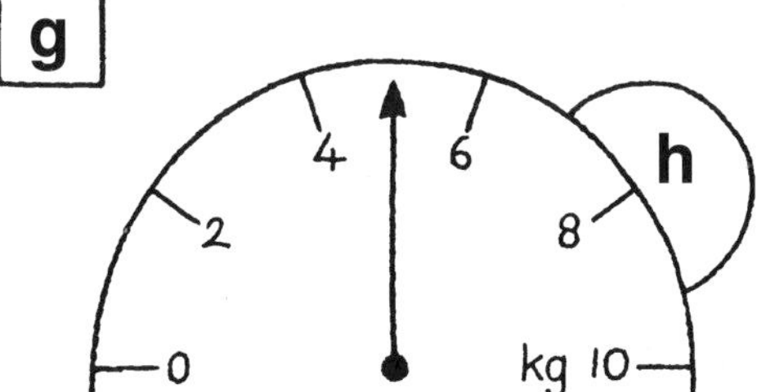

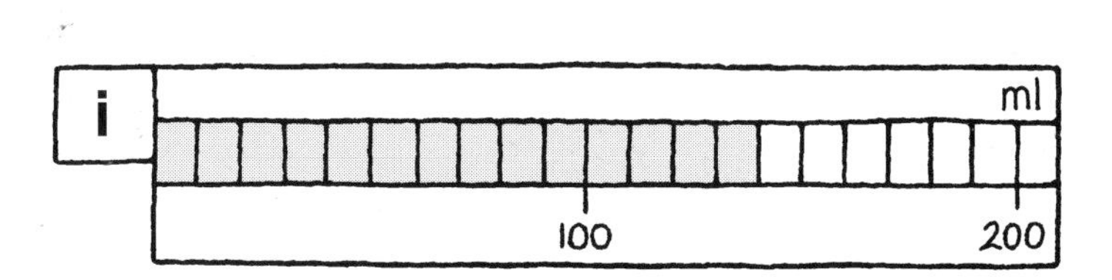

CONTROL PANEL READINGS

a 10 ml	b ______	c ______
d ______	e ______	f ______
g ______	h ______	i ______

• **Write which dials show readings of:**

mass | length | capacity

Teachers' note Revise the abbreviations cm, m, km, kg, g and ml, and tell the children to write the unit after each reading. Remind them always to look at the numbers and units on a scale to work out how much each interval is worth.

Game on

• **Play this game with a partner.**

☆ You need counters in two colours.
☆ Take turns to choose a square.
☆ Name an item of the length, mass or capacity shown on the square. If your partner agrees, put one of your counters on the square.
☆ The winner is the first to get four in a line.

The pictures below might help you.

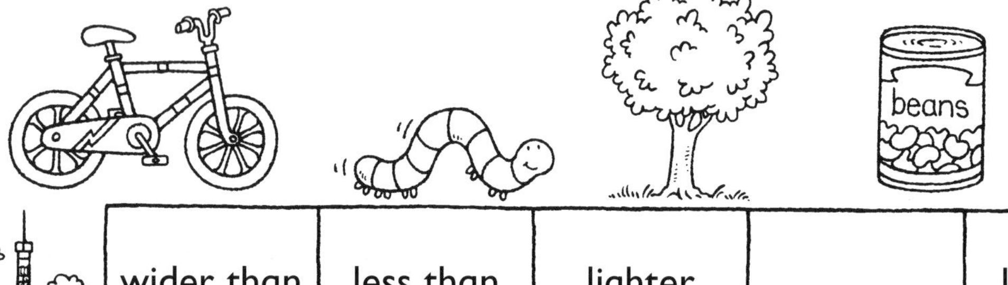

wider than $\frac{1}{2}$ m	less than 25 cm	lighter than 10 g	about 1 cm	less than 5 mm
less than 1 kg	less than 100 ml	about 1 kg	further than 10 km	more than 1 kg
heavier than 2 kg	smaller than 1 m	narrower than 10 cm	more than 1000 ml	about 1 mm
less than 1 litre	about 150 g	about $\frac{1}{2}$ kg	over 1 km	about 1 litre
about 10 cm	wider than 2 m	lighter than 500 g	taller than 1 m	about 40 m

Teachers' note Show the children a centimetre cube and remind them that if the cube were filled with water, it would hold 1 millilitre and would weigh 1 gram. Then show them a 1000 Dienes block to demonstrate 1 litre and 1 kilogram in the same way. Ask the children to write down any amounts or items they disagree on. Discuss these in the plenary.

Snail trails

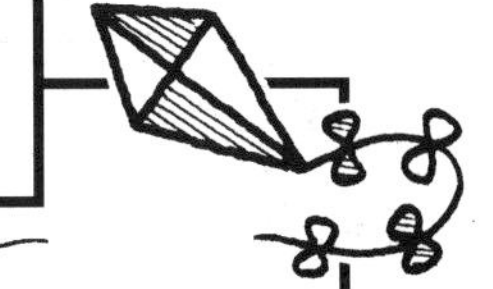

A snail has left a slime trail around the edge of each shape.

- **Find the length of each trail.**

This is called the perimeter.

1.

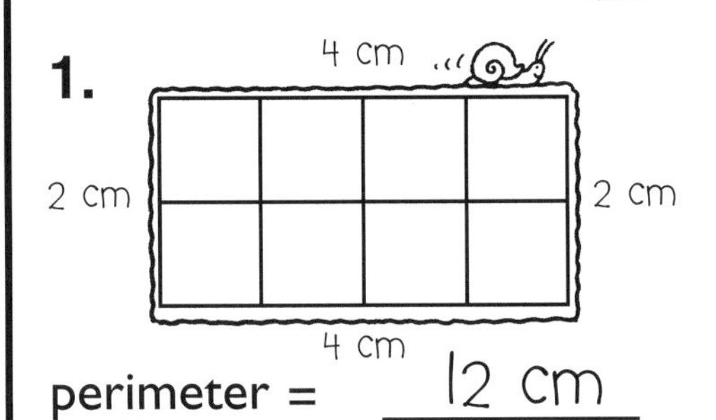

perimeter = 12 cm

2.

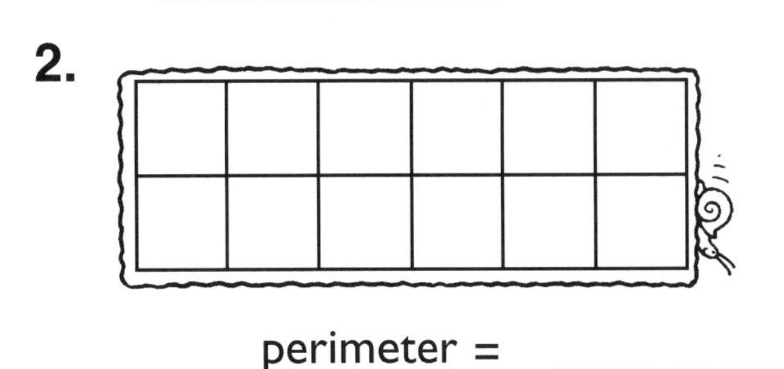

perimeter = ________

3.

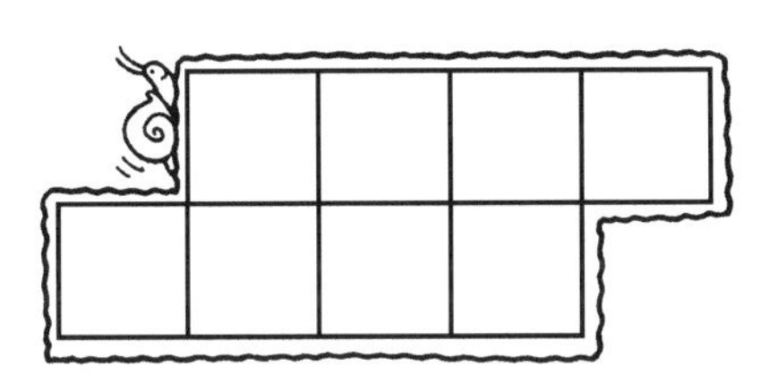

perimeter = ________

4.

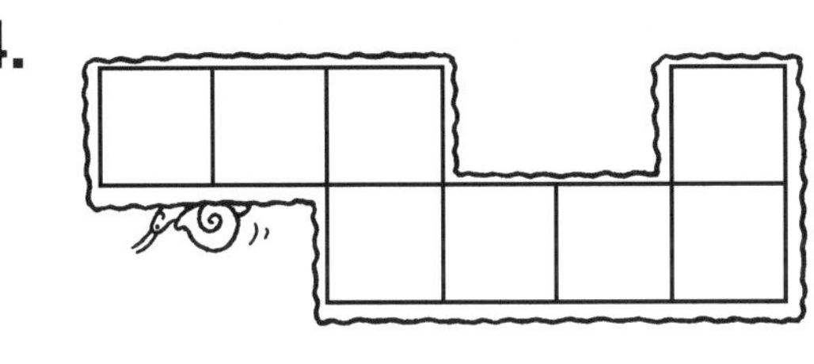

perimeter = ________

5.

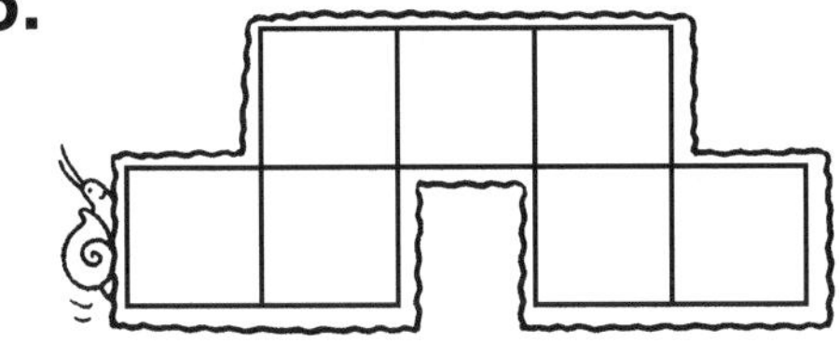

perimeter = ________

6.

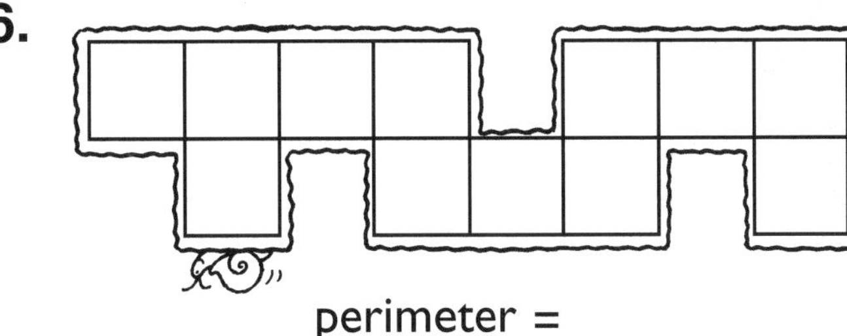

perimeter = ________

7.

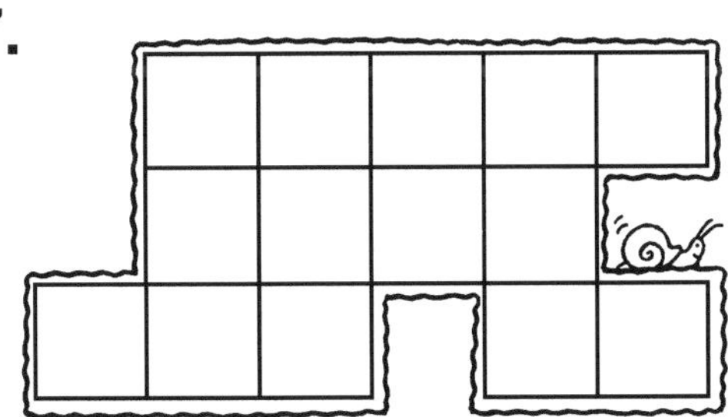

perimeter = ________

8.

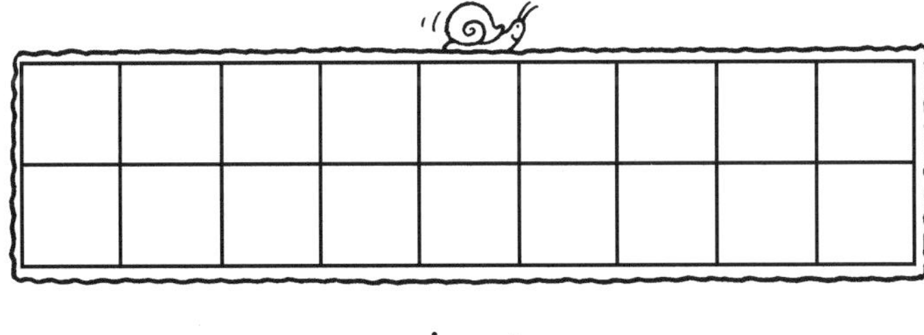

perimeter = ________

- **On squared paper, draw five different shapes with a** perimeter **of 12 cm.**

Teachers' note Children often confuse perimeter and area and incorrectly count squares when measuring perimeter. Remind them that perimeter is a length, measured in centimetres. Encourage the children to jot down parts of the perimeter as they go, then total them. Provide 1 cm squared paper for the extension activity.

Perimeter patterns

Jo has drawn some shapes on 1 cm triangular dotted paper.

• Write the perimeter of each shape.

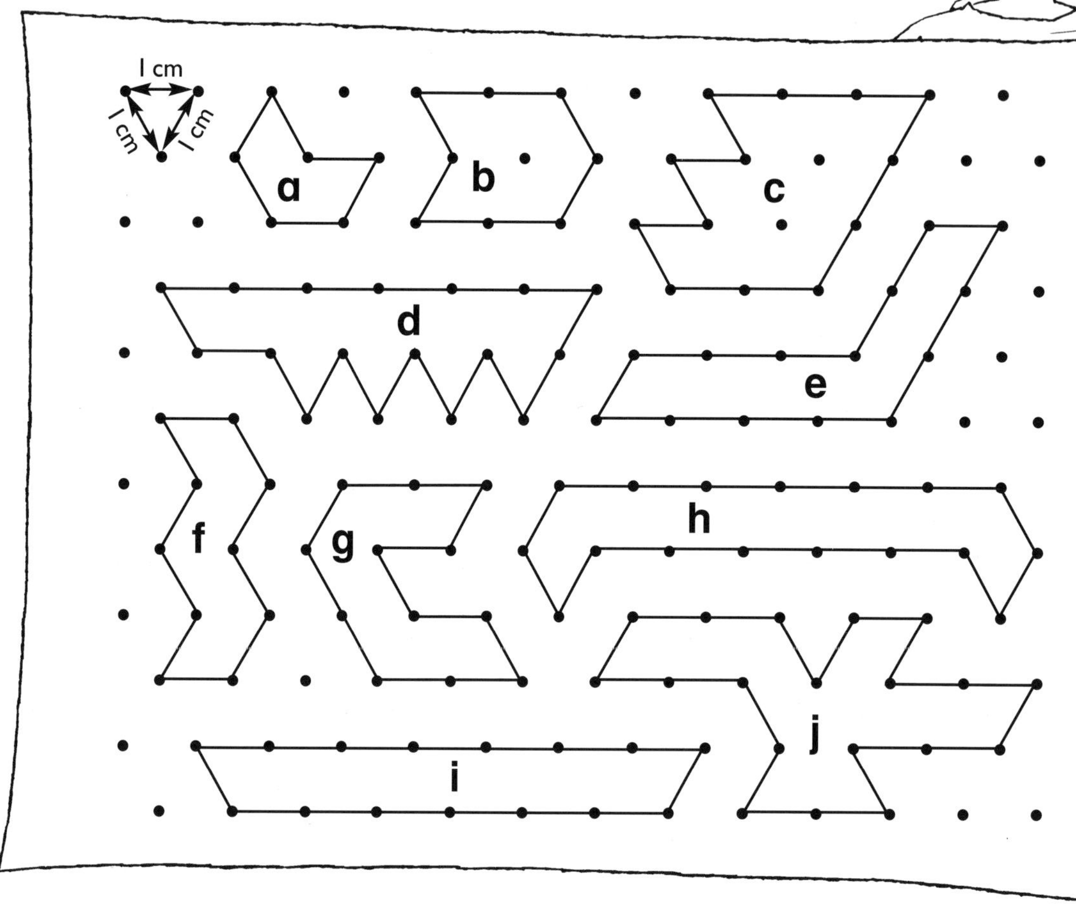

a 6 cm b ______ c ______ d ______ e ______

f ______ g ______ h ______ i ______ j ______

Now try this!

• Draw four more shapes on this paper.

• Find the perimeter of each shape.

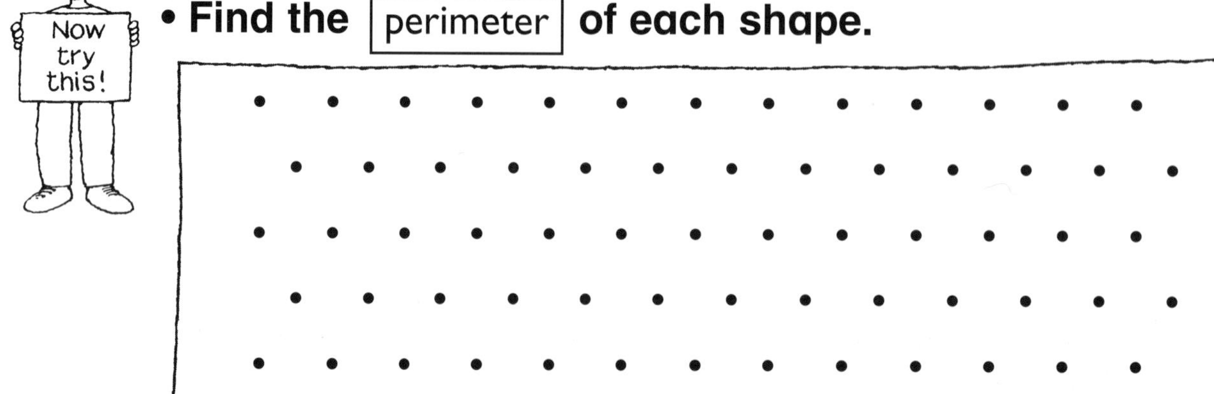

Teachers' note Encourage the children to jot down parts of the perimeter as they go, then total them.

The dolls' house

The dolls' house needs new carpets.

- Look at the sketches. They are on 1 cm squared paper.
- Write the area of each carpet.

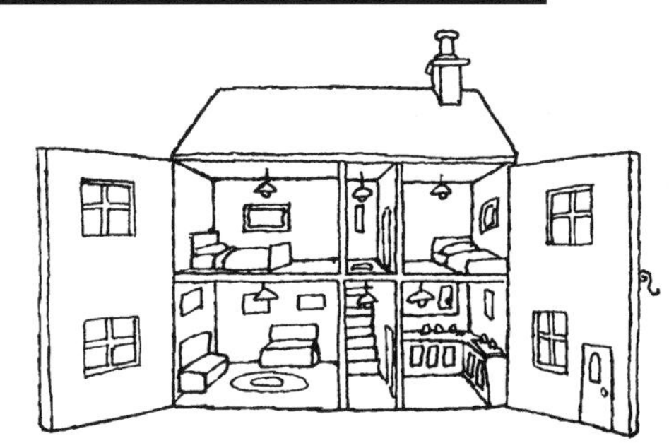

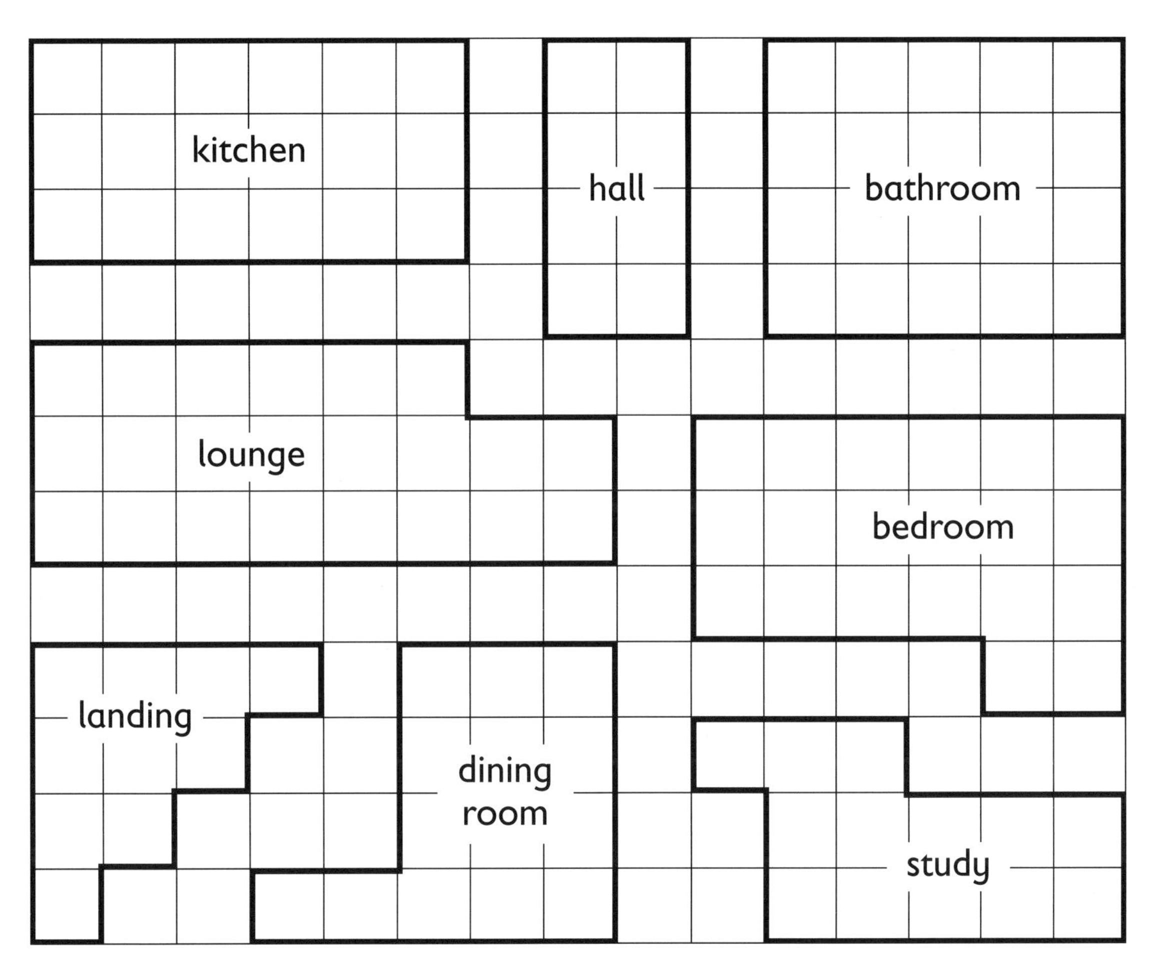

1. kitchen 18 cm²
2. hall ____ cm²
3. bathroom ____ cm²
4. lounge ____ cm²
5. bedroom ____ cm²
6. landing ____ cm²
7. dining room ____ cm²
8. study ____ cm²

Another piece of carpet has an area of 17 cm².

- On squared paper, draw three different shapes it might be.

Teachers' note Discuss quicker ways of finding the areas of rectangles, such as the kitchen, hall and bathroom carpets, without counting each square. Provide the children with 1 cm squared paper for the extension activity.

Brilliant badges

- **Colour a pattern on each badge. Match the areas shown.**

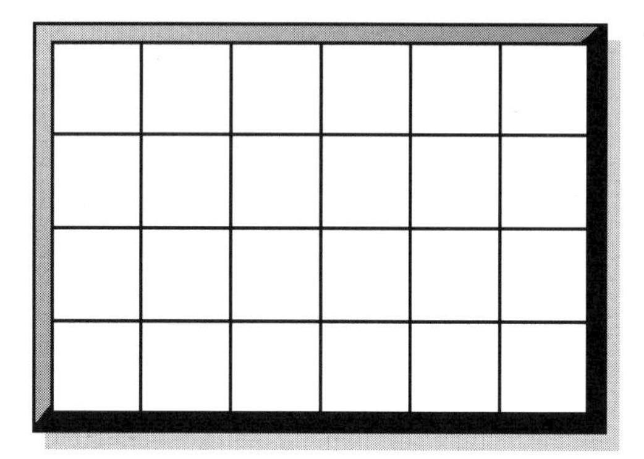

red 6 cm^2
blue 8 cm^2
yellow 10 cm^2

You can colour half squares and whole squares.

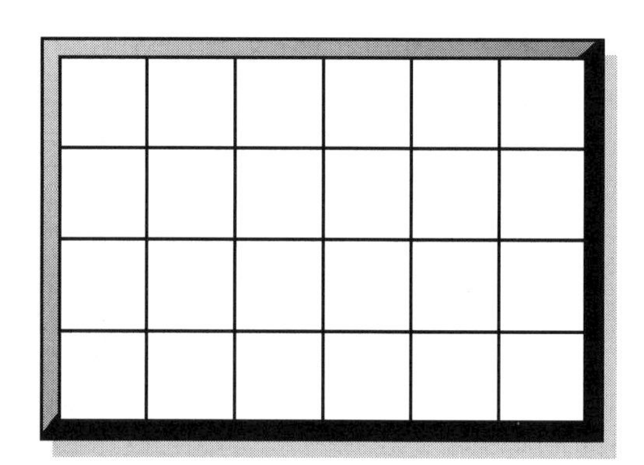

green 7 cm^2
white 2 cm^2
yellow 6 cm^2
purple 9 cm^2

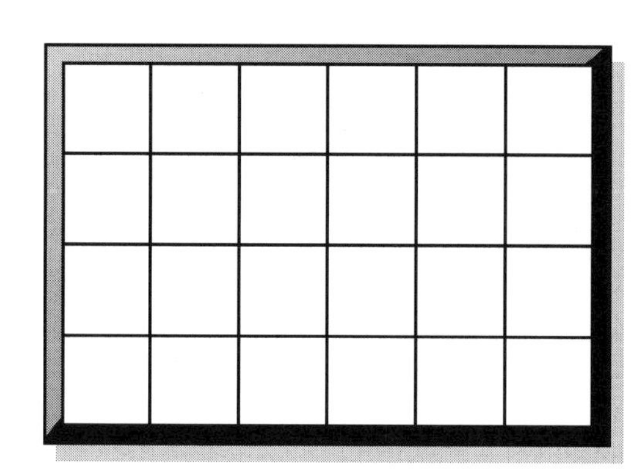

red $6\frac{1}{2}$ cm^2
blue 4 cm^2
orange $5\frac{1}{2}$ cm^2
yellow 8 cm^2

Now try this!

- **Write the area of this badge that is:**

grey ________
striped ________
black ________
white ________

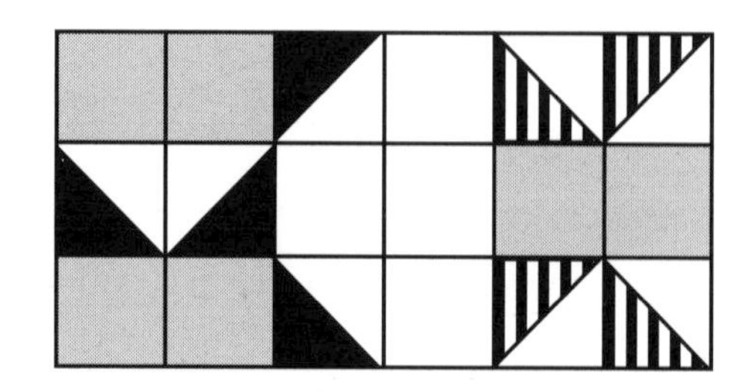

- **Check that your answers total 18 cm^2.**

Teachers' note The children will need a range of coloured pencils for this activity. Encourage them to check the total of the areas coloured. They could make their own badges of different shapes and sizes on squared paper.

Farmer's fields

Farmer Evans splits each field in half, using one fence.

- **Draw as many different ways as you can find. The halves should not look the same.**

Remember, I split the **area** of the field in half.

Teachers' note Encourage the children to draw fences along the lines shown and to develop a method of finding different ways. As an extension activity, the children could be asked to find further ways using diagonal lines. They should group any ways that are reflections or rotations of others.

Magic months

Amy uses her knuckles to help her remember how many days are in each month.

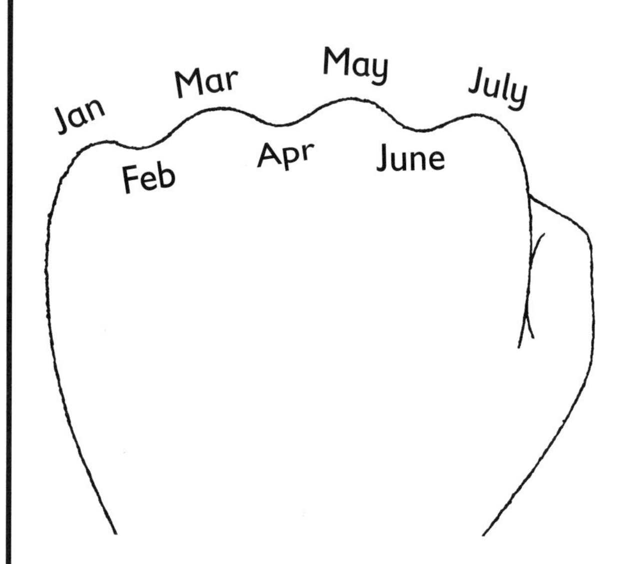

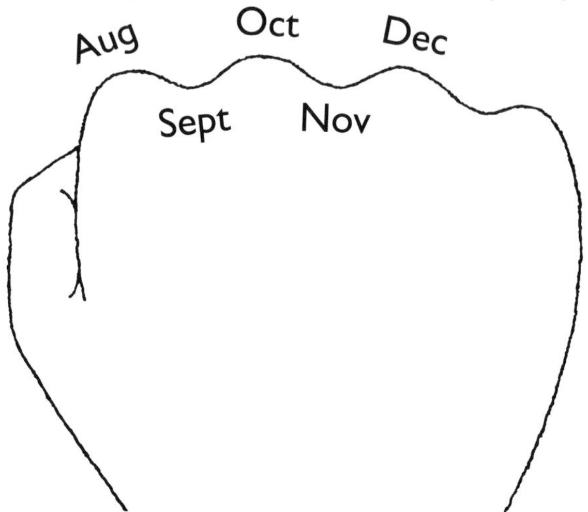

- **Use this, or another way, to find the number of days in:**

1. July 31
2. November _____
3. March _____
4. August _____
5. December _____
6. September _____
7. June _____
8. April _____
9. February _____ or _____
10. October _____
11. May _____
12. January _____

- **How many days altogether in:**

13. January and February? _____ or _____
14. March and April? _____
15. May and June? _____
16. July and August? _____

- **Which date is exactly in the middle of the year?**

____

Teachers' note Demonstrate to the class how to use your knuckles to find the months that have 31 days. Explain that all the months between the knuckles have 30 days except for February, which has 28, and 29 in a leap year. Children should know the rhyme '30 days hath September' before starting this activity.

More or less?

- **Write your date of birth.** ______________________
- **Write your age in years, months and days.**

Example: 8 years, 4 months and 17 days.

____

- **Tick whether you are** more **or** less **than these ages.**

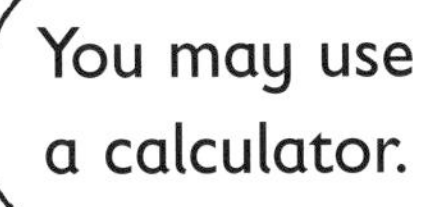
You may use a calculator.

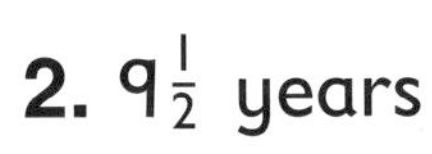

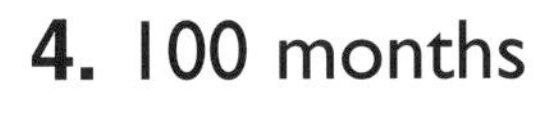

1. $8\frac{1}{2}$ years	more	☐	less	☐
2. $9\frac{1}{2}$ years	more	☐	less	☐
3. 90 months	more	☐	less	☐
4. 100 months	more	☐	less	☐
5. 400 weeks	more	☐	less	☐
6. 500 weeks	more	☐	less	☐
7. 3 000 days	more	☐	less	☐
8. 3 400 days	more	☐	less	☐
9. 80 000 hours	more	☐	less	☐
10. 70 000 hours	more	☐	less	☐

Teachers' note Some children may need to assume that each month is 30 days and each year is 365 days. More able children could be more accurate and may wish to consider leap years. Remind the children of the relationships between the units of time: 12 months/52 weeks/ 365(6) days = 1 year; 24 hours = 1 day.

Day or night?

- Write whether these are day or night times.
- Write what you might be doing at the times.

11:07 pm	This is at night-time. I am usually asleep.
9:36 am	
1:11 pm	
7:57 am	
4:17 pm	
11:29 am	
7:36 pm	
10:01 am	

Teachers' note Prepare for this activity by asking the children to keep a diary for a day, noting the times that key events happen. Encourage them to consider the start and end times of lessons. As an extension activity, a timeline from midnight to midnight can be drawn showing these times.

Good timing

- **Write these times in words.**

seventeen minutes
past seven

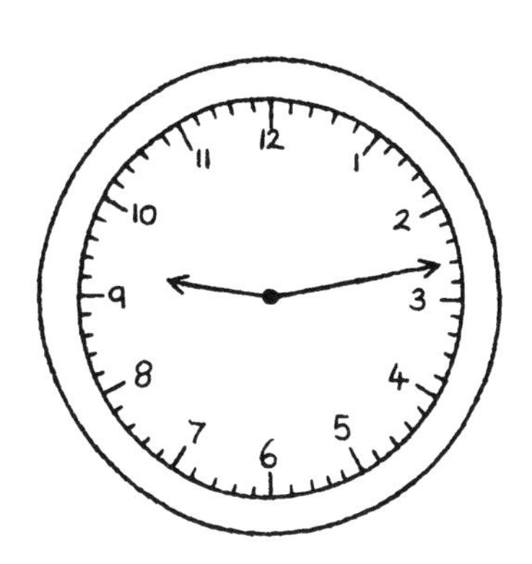

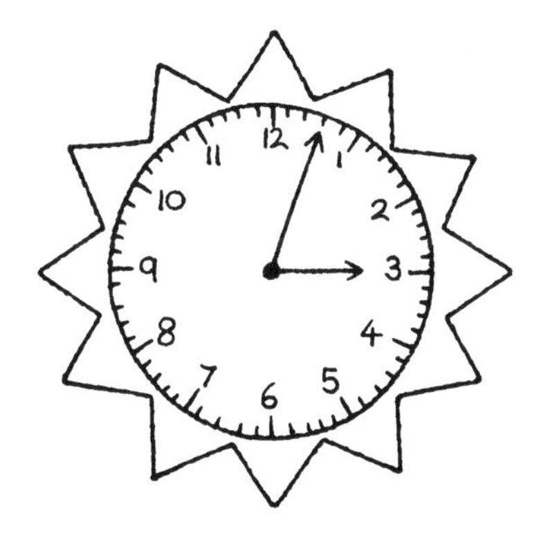

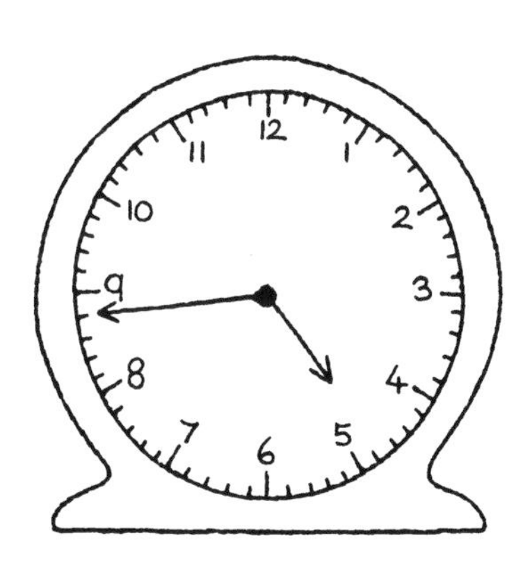

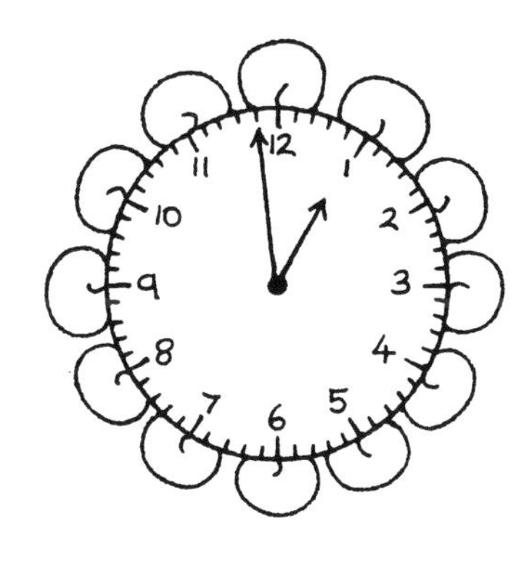

- **Write each time in** digital **form.** Example: 7:17

Teachers' note The hands on the clocks can be masked to provide a flexible resource.

Watch it!

- **Write these times in words.**

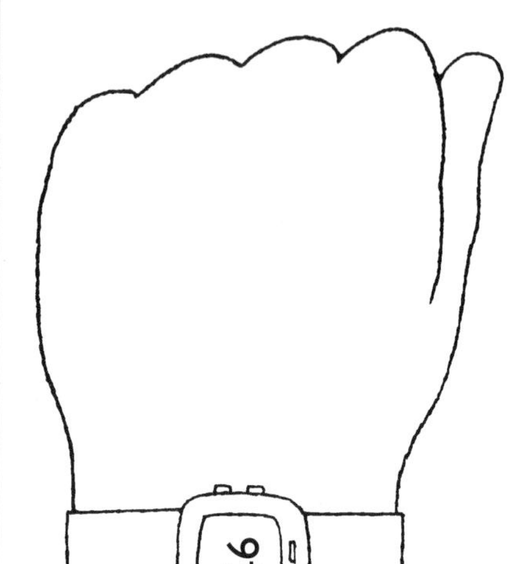

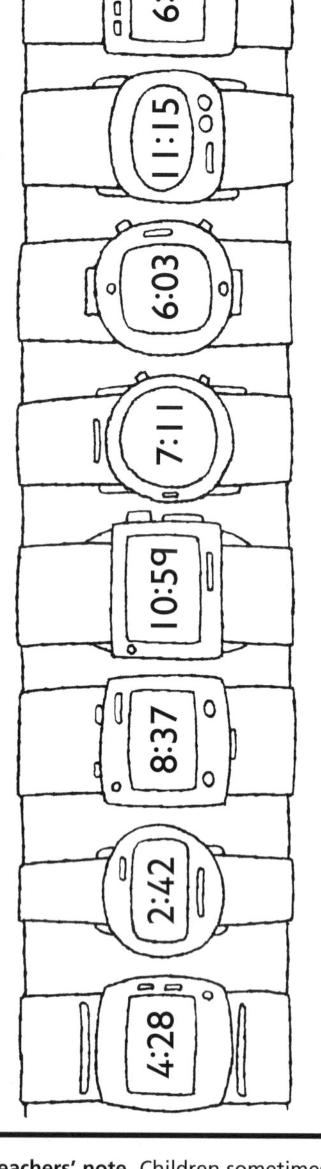

fourteen minutes to nine

Teachers' note Children sometimes write times such as 6:30 as 'six thirty' without realising it is half past six or 30 minutes past six. Ensure that answers are written in full. Some children may need a clock with moveable hands or a clock face stamp to help them count the minutes.

Time travel

Here is the airbus timetable on Planet Zogo.

	Airbus 1	Airbus 2	Airbus 3	Airbus 4	Airbus 5	Airbus 6
Gravity Pods	8:30	9:15	9:45	10:20		
Crater Cafe	8:50	9:35	10:05			
Hologram Hall	9:10	9:55	10:25		12:05	
Moon Deck	9:30	10:15	10:45			1:57

- **Write the time it takes to travel between these stops.**

1. Crater Cafe → mins → Hologram Hall

2. Crater Cafe → mins → Moon Deck

3. Gravity Pods → mins → Hologram Hall

4. Gravity Pods → mins → Moon Deck

- **Complete the timetable.**

Now try this!

You are at the Gravity Pods. You need to be at Crater Cafe just before 9:40, then travel on to the Moon Deck for 11:30.

- **Write the airbuses you need.**

Airbus ☐ at ☐. Then Airbus ☐ at ☐.

Crater Cafe

Teachers' note Some children may need help bridging the hour. Encourage them to find the time gap before the hour and then add it to the time gap after the hour. Some children may need help identifying that there are 20 minutes between each stop in order to complete the timetable.

Today's the day

• **Answer the questions. Use this year's calendar.**

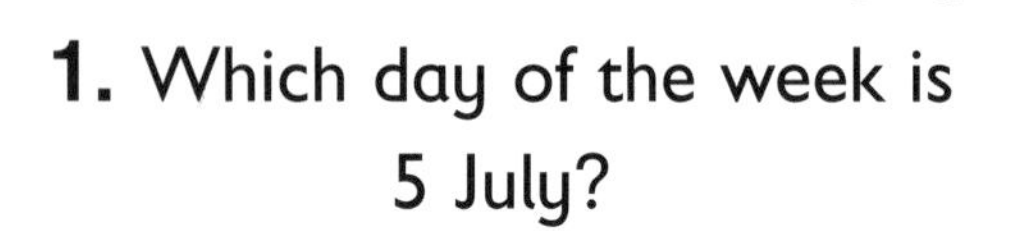

1. Which day of the week is 5 July?

____

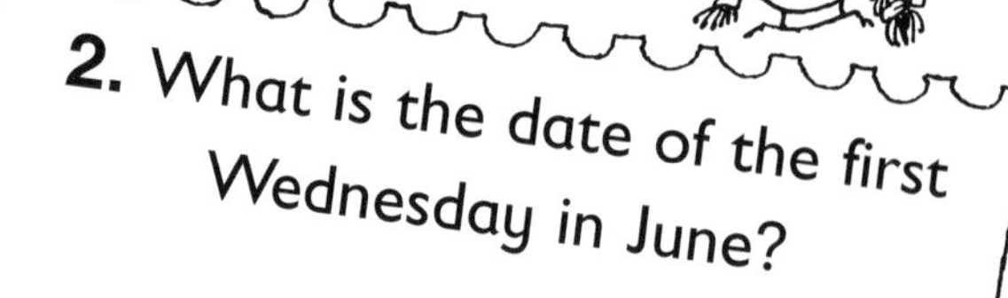

2. What is the date of the first Wednesday in June?

____

3. How many days are there from 26 February to 11 March?

______

4. What is the date of the third Friday in September?

____

5. Which date is exactly four weeks from today?

____

6. Which day of the week is 25 December?

____

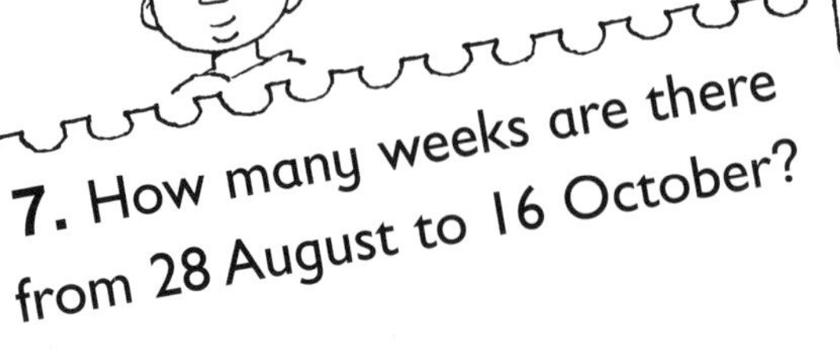

7. How many weeks are there from 28 August to 16 October?

______

8. Write the dates of all the Fridays in February.

____

____

• **Write the date and day of the** 23rd **of each month.**

Teachers' note Questions could be masked and altered to provide a wider range of questions to solve. When the children are counting the number of days from one date to another, encourage them to count on from the date (i.e. not include it in the count).

Prisms and pyramids

- **Find the number of**  edges **of each prism.**

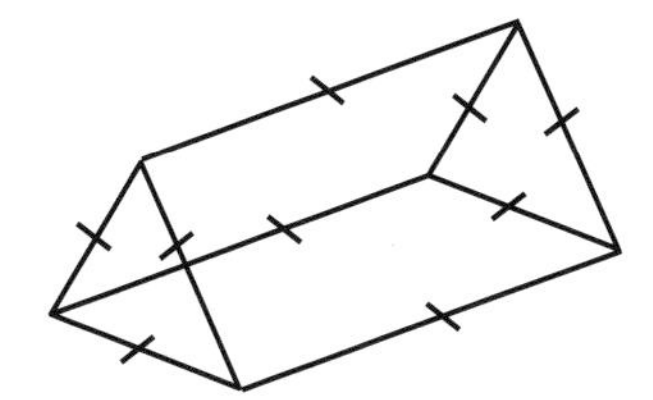

A **triangular** prism

has 9 edges.

Mark each edge as you count it.

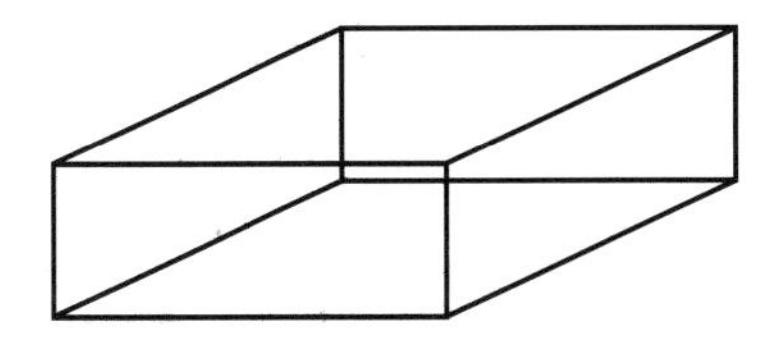

A **rectangular** prism

has _____ edges.

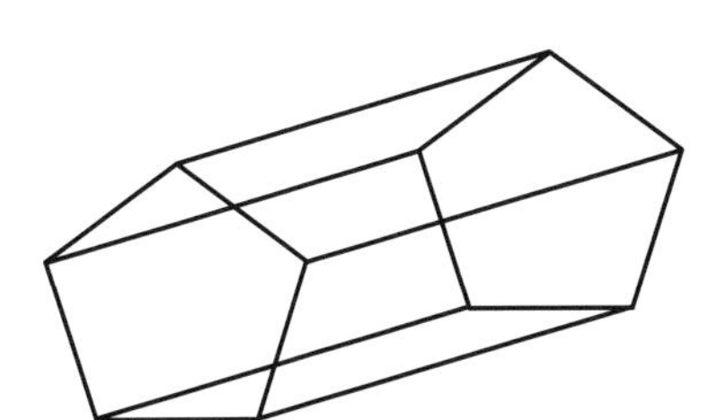

A **pentagonal** prism

has _____ edges.

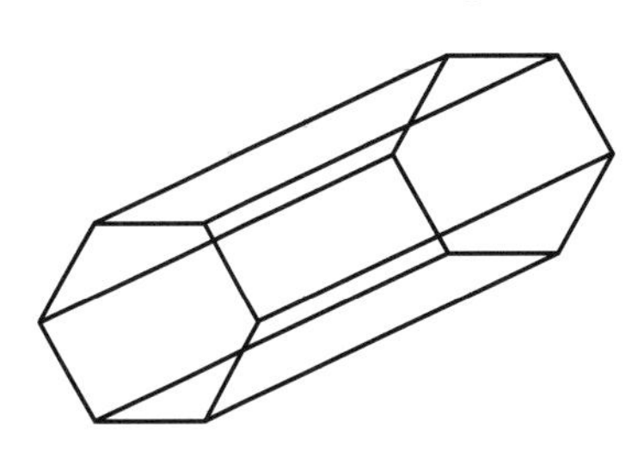

A **hexagonal** prism

has _____ edges.

- **Find the number of**  edges **of each pyramid.**

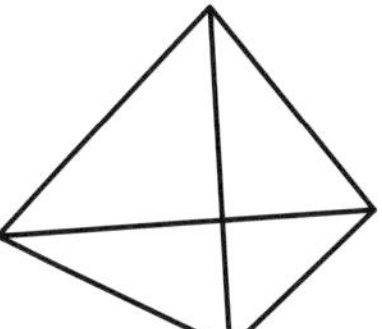

A **triangular-based** pyramid

has _____ edges.

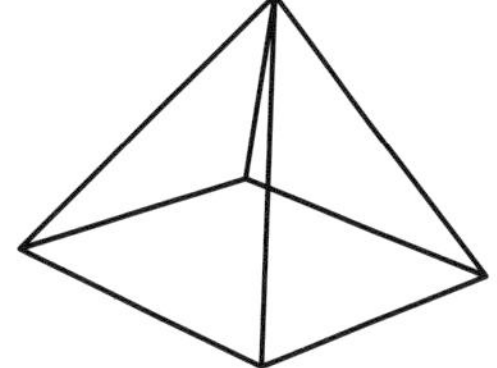

A **square-based** pyramid

has _____ edges.

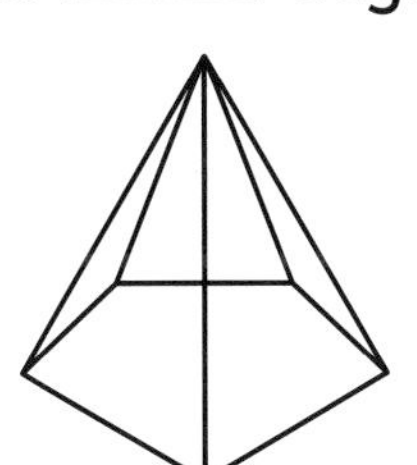

A **pentagonal-based** pyramid

has _____ edges.

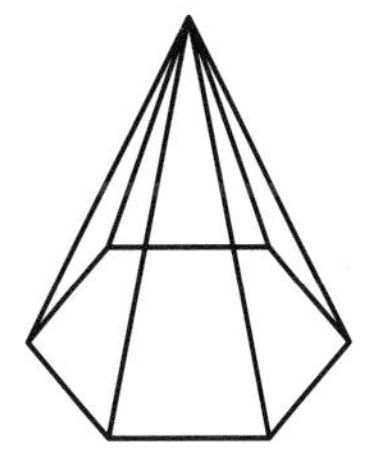

A **hexagonal-based** pyramid

has _____ edges.

- **What do you notice about**

the number of edges of a prism?

the number of edges of a pyramid?

Teachers' note Some children may need a set of solid shapes to refer to. In the extension activity, encourage them to find other prisms and pyramids to test their hypotheses. For reference: a prism has the same cross-section all along its length, and identical end faces; a pyramid has a polygon for its base, and triangular faces which meet at one vertex.

Match and describe

- **Join each shape to its name.**
- **Write two things about each shape.**

Think about faces, vertices and edges.

Name	Description
cube	six identical faces
cylinder	
tetrahedron	
triangular prism	
square-based pyramid	

Now try this!

- **Write the shape which has:**

the fewest faces ____________________

the most vertices ____________________

exactly eight edges ____________________

Teachers' note When children are writing their descriptions, encourage them to make reference to the number of faces, vertices and edges, and the shapes of the faces. As a further extension, the children could write the number of faces, vertices and edges for each shape.

Finders keepers

- **Play this game with a partner.**

☆ You need shapes, a dice and two counters.
☆ Take turns to roll the dice and move your counter.
☆ Find a shape that fits the description you land on, if you can. Keep the shape.
☆ The winner is the player with the most shapes.

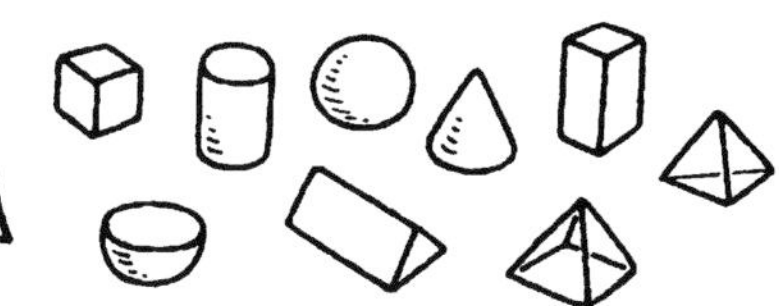

start

It has one face and no vertices.

It has two triangular faces.

All its faces are identical.

Some of its faces are rectangles.

It has one square face.

It has eight vertices.

It has nine edges.

It has one curved face and one flat face.

It has only one vertex.

At least two of its faces are triangles.

It has only one edge.

It has two more edges than faces.

It has the same number of faces as vertices.

Its end faces are identical.

It has more edges than vertices.

It has only flat faces.

At least one of its faces is curved.

None of its faces is curved.

It has no vertices.

finish

Teachers' note Before the children begin the activity, discuss how to count the number of edges of shapes with curved faces. Provide one set of the following shapes: cube, cuboid, cylinder, sphere, cone, hemi-sphere, triangular prism, square-based pyramid, tetrahedron. The children should check each other's selection at each turn.

Model making

- **Make each model using interlocking cubes.**
- **Write the number of cubes needed.**

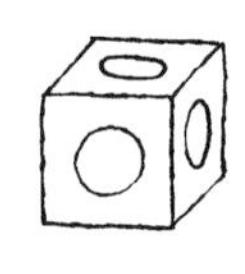

1\.

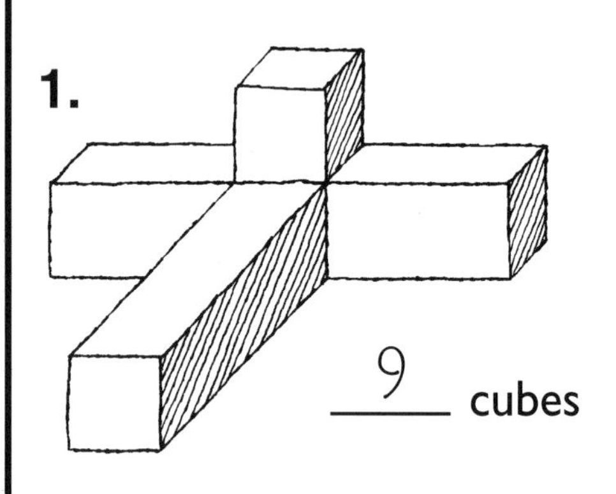

9 cubes

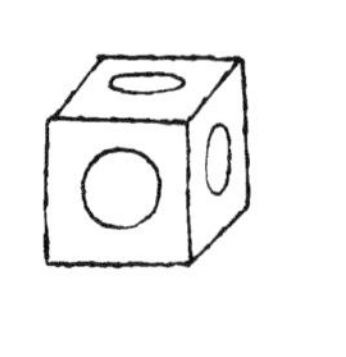

2\.

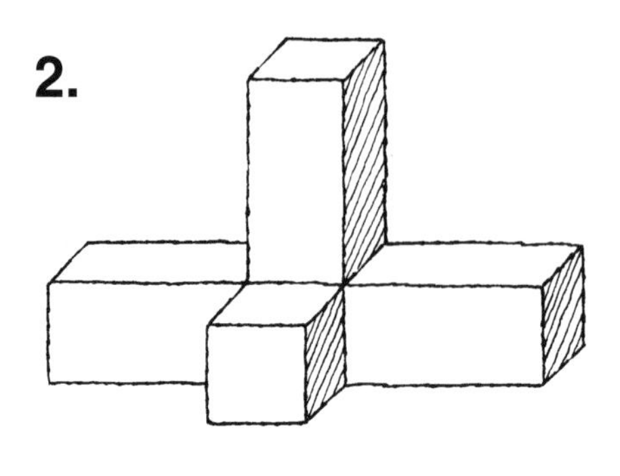

_____ cubes

3\.

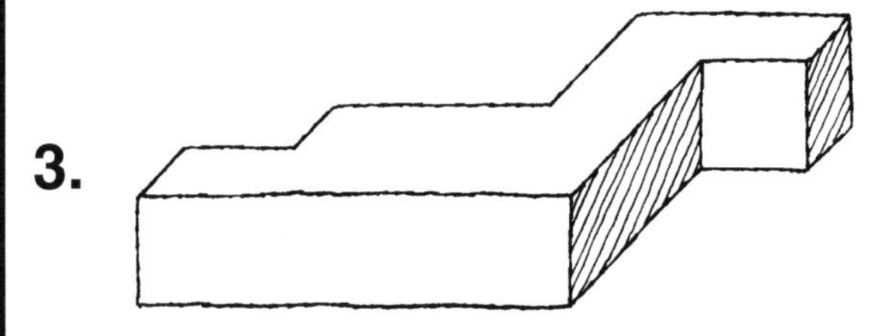

_____ cubes

4\.

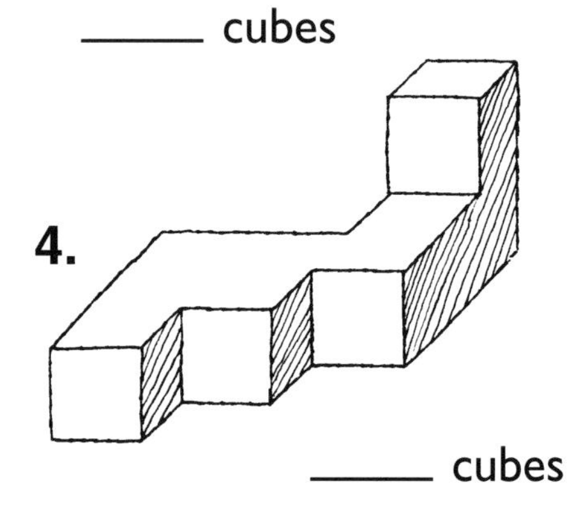

_____ cubes

5\.

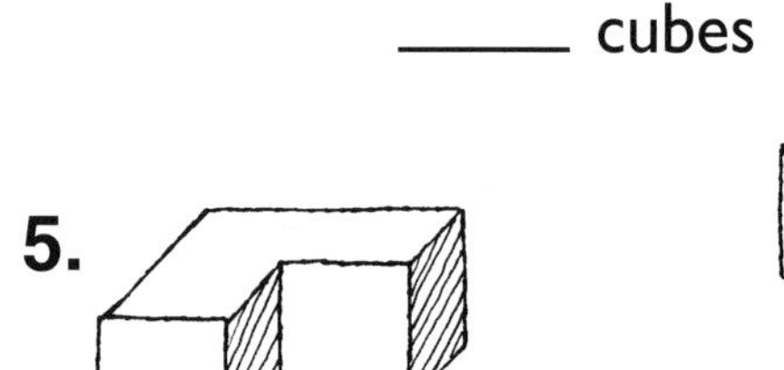

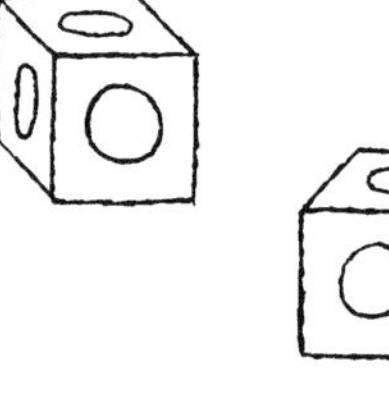

_____ cubes

6\.

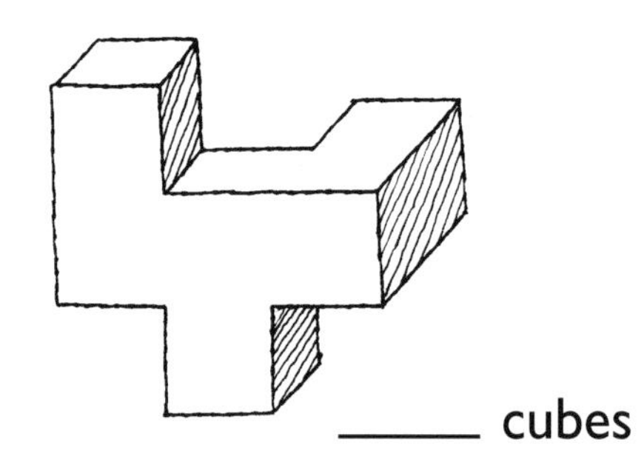

_____ cubes

7\.

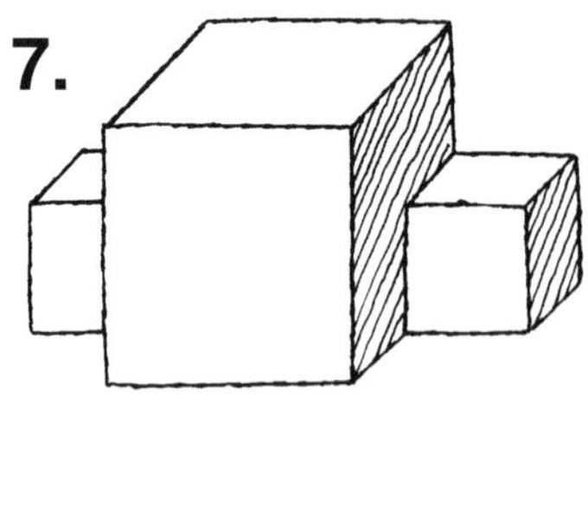

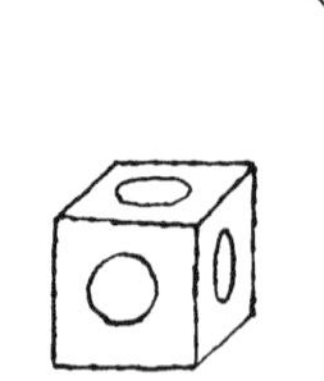

_____ cubes

8\.

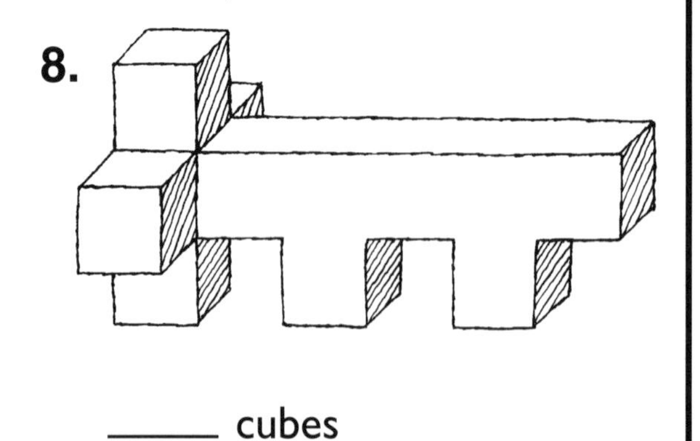

_____ cubes

- **Make as many different models as you can using eight cubes.**

Teachers' note Encourage the children to work in pairs, both making the same shape and then comparing them.

Fishing nets

You can fold these nets **to make solid shapes.**

- **Which solid shape will each net make?**

1.

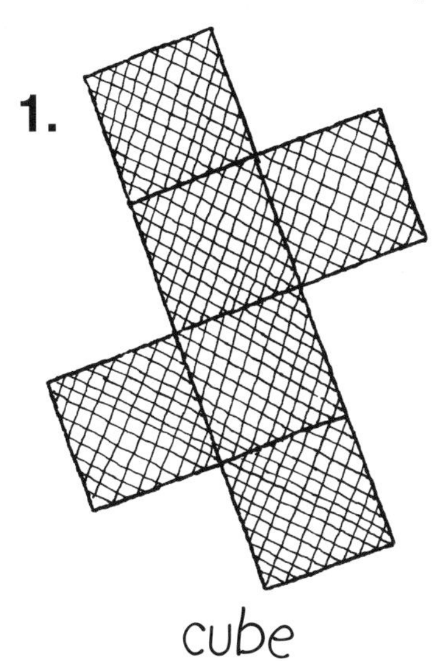

cube
____

2.

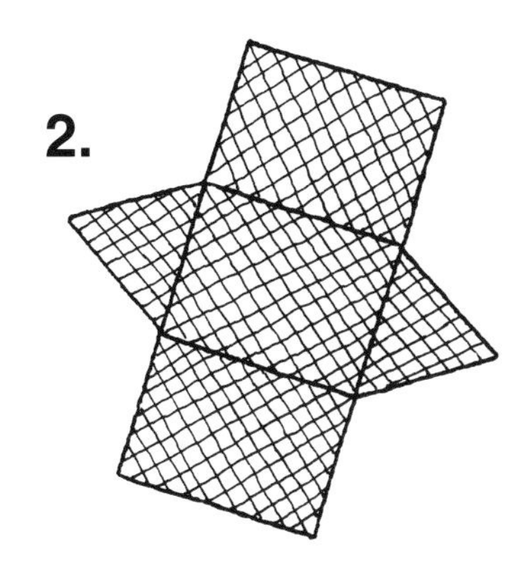

____

3.

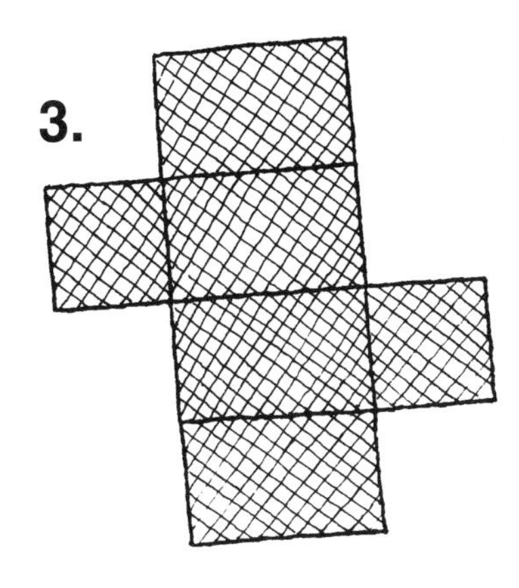

____

4.

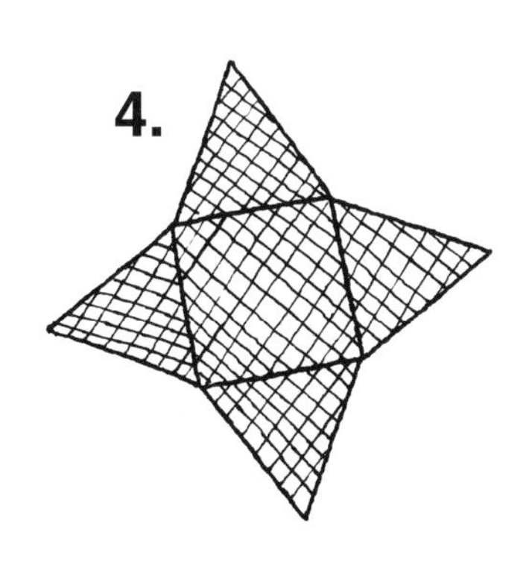

____

5.

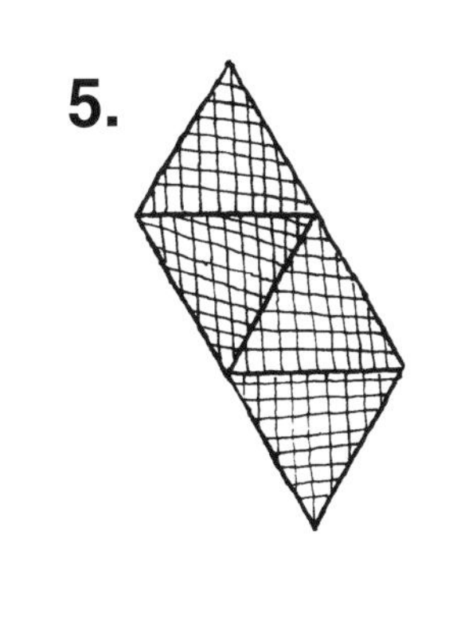

____

6.

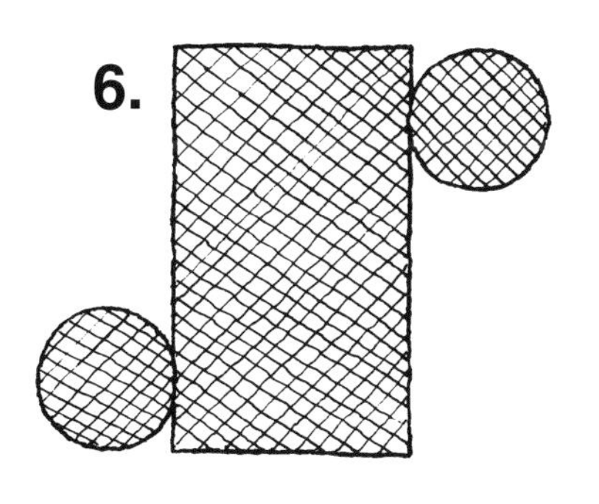

____

7.

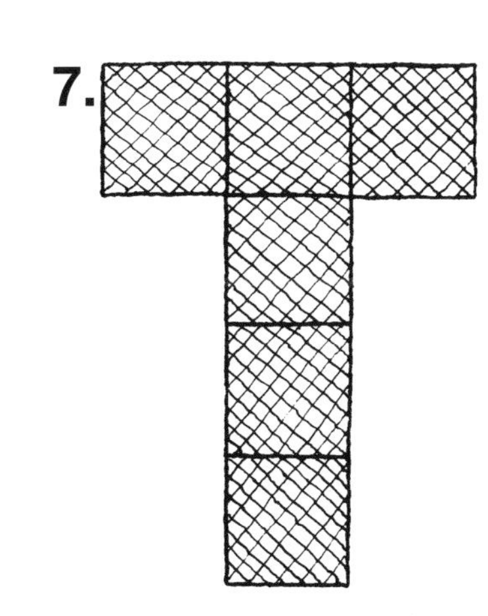

____

- **On squared paper, draw nets for a cube.
  Find as many different ways as you can.**

Teachers' note Ensure that the children appreciate that the same solid shape can have many different nets. Some children may need construction material such as Polydron to help them with this activity.

Zip, Zap or Zop?

On the planet Hepta, 7-sided shapes are worth 1 million pounds. Other shapes are worth nothing.

- Colour the heptagons.
- Which alien's shapes are worth the most? ________

A 7-sided shape is called a **heptagon**.
Zip
Zap
Zop

Now try this!

- If hexagons are worth £50 000 and octagons are worth £10 000, which alien's shapes are worth the most? ________

This time a heptagon is worth nothing.

Teachers' note Different values can be given to each shape, for example, hexagons £10, heptagons £100, octagons £1000, and the total value calculated.

Jump to Jack

Jack and his friends are playing a game. Jack's friends have to jump from shape to shape to the other side of the mat.

- **Draw the routes. Use a different colour for each friend.**

You can jump vertically, horizontally or diagonally.
Jack

I can only jump on shapes which have **more than five sides.**
I can only jump on shapes which have **at least one right angle.**
I can only jump on **symmetrical** shapes.
I can only jump on **quadrilaterals.**

Now try this!

Sally can only jump on regular **shapes.**

- **Draw a route for Sally.**

Teachers' note The children can make their own Jump to Jack 'mat' using shapes made from sticky paper. These could be used as a lively display.

Concave or convex?

This shape is concave. This shape is convex.

- **Colour concave shapes green and convex shapes yellow.**

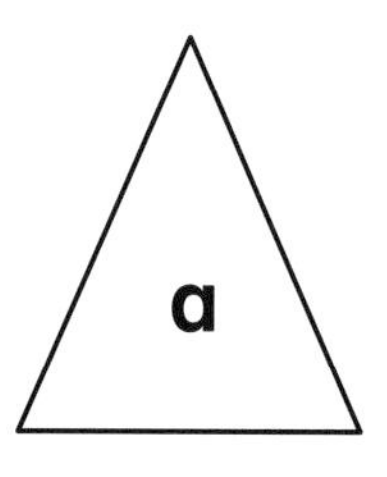

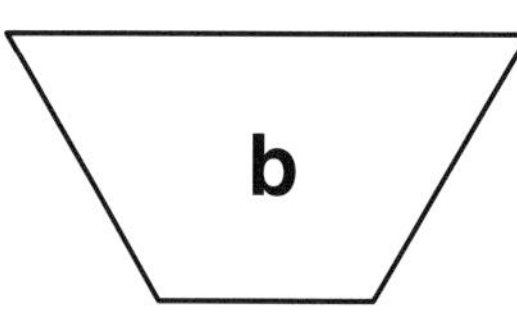

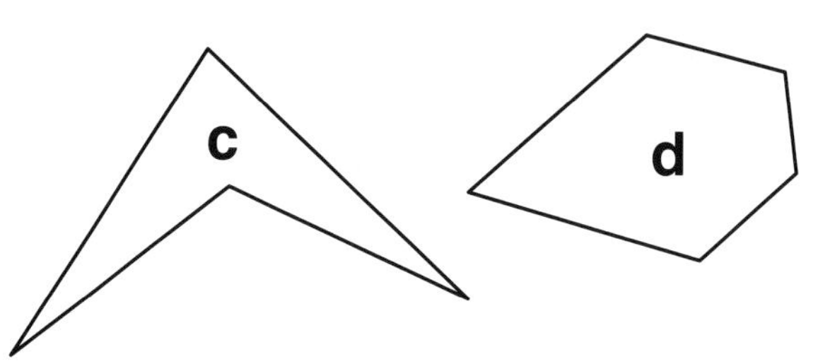

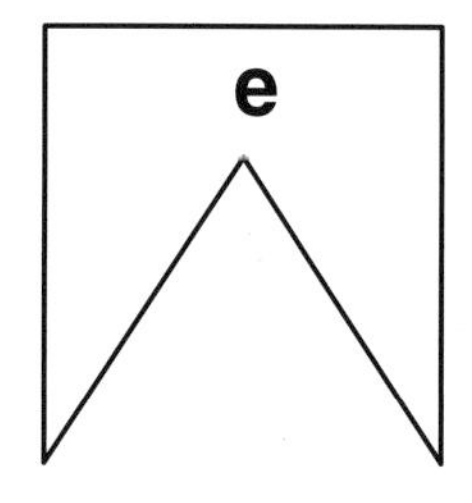

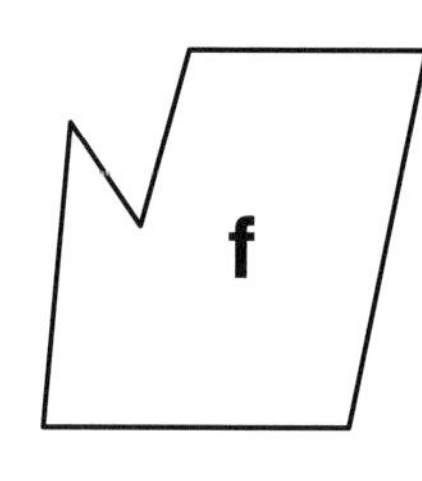

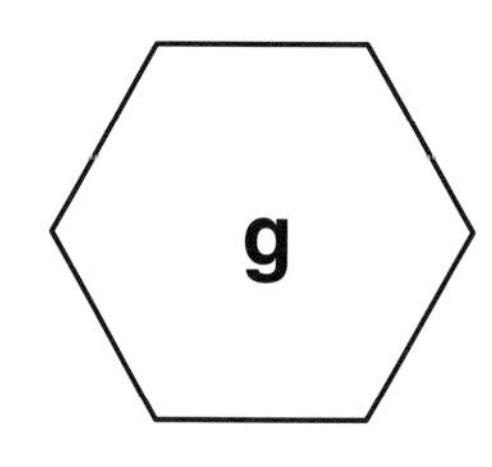

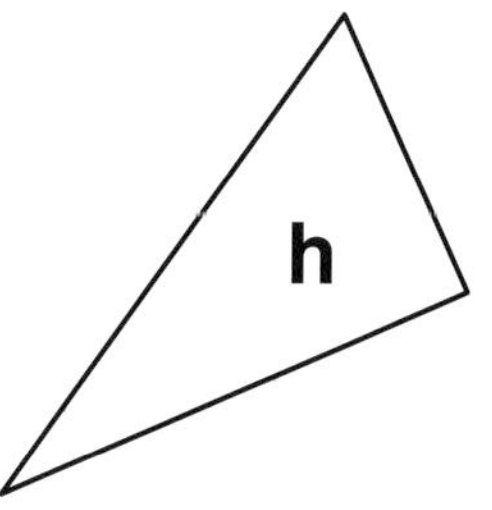

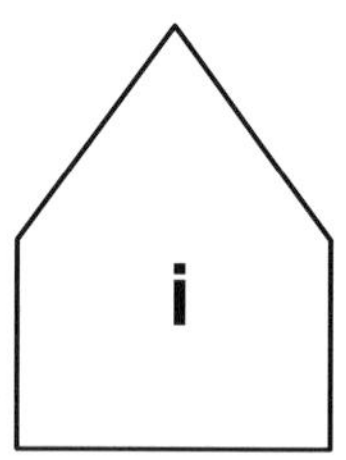

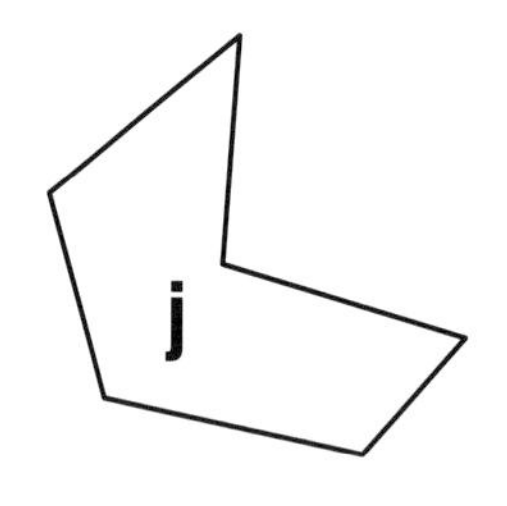

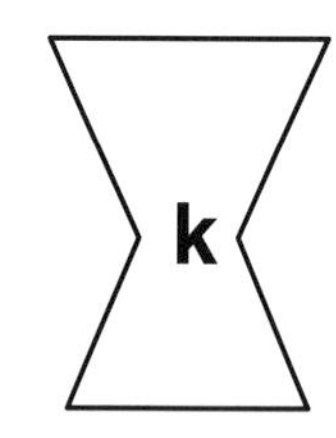

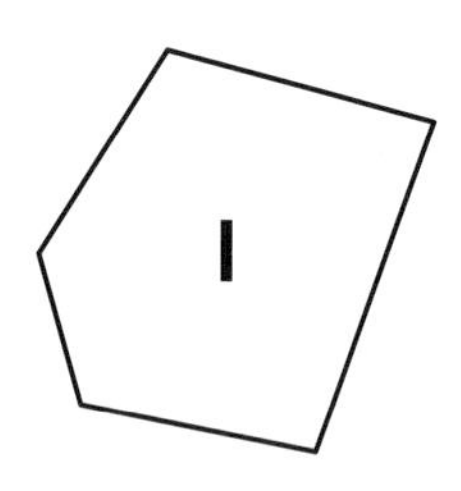

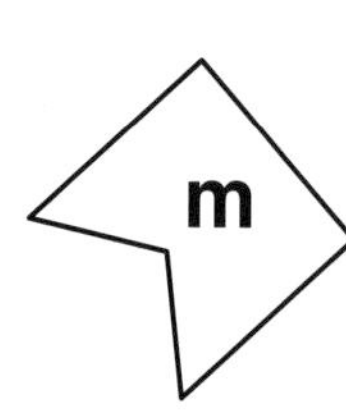

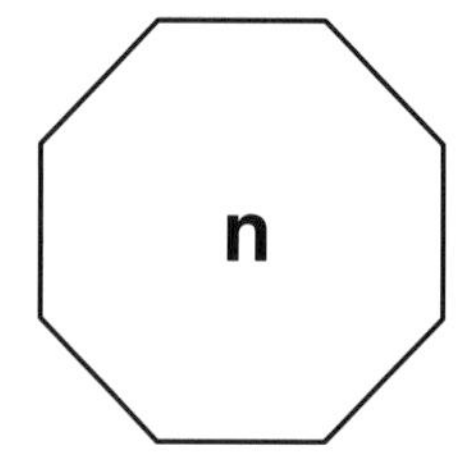

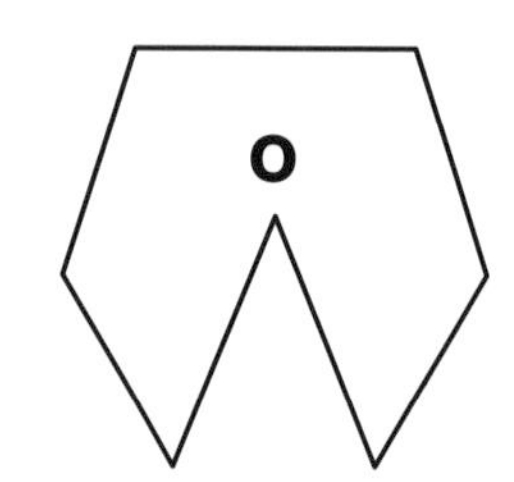

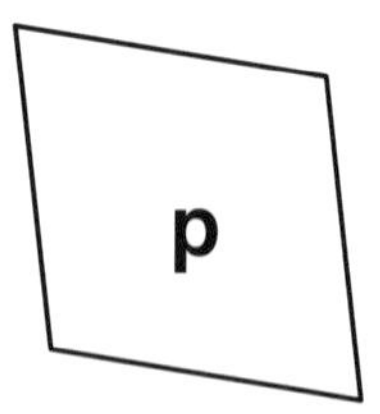

- **Draw a robot using five concave shapes and three convex shapes.**

Teachers' note Ensure that the children understand the terms 'concave' and 'convex' before they attempt this activity. Point out the word 'cave' in 'concave' and encourage them to think of caves pointing inwards.

Trying triangles!

This pattern is made from triangles.

- **Colour** isosceles **triangles blue and** equilateral **triangles red.**

An **isosceles** triangle has two sides that are equal and two angles that are equal.

An **equilateral** triangle has three sides that are equal and three angles that are equal.

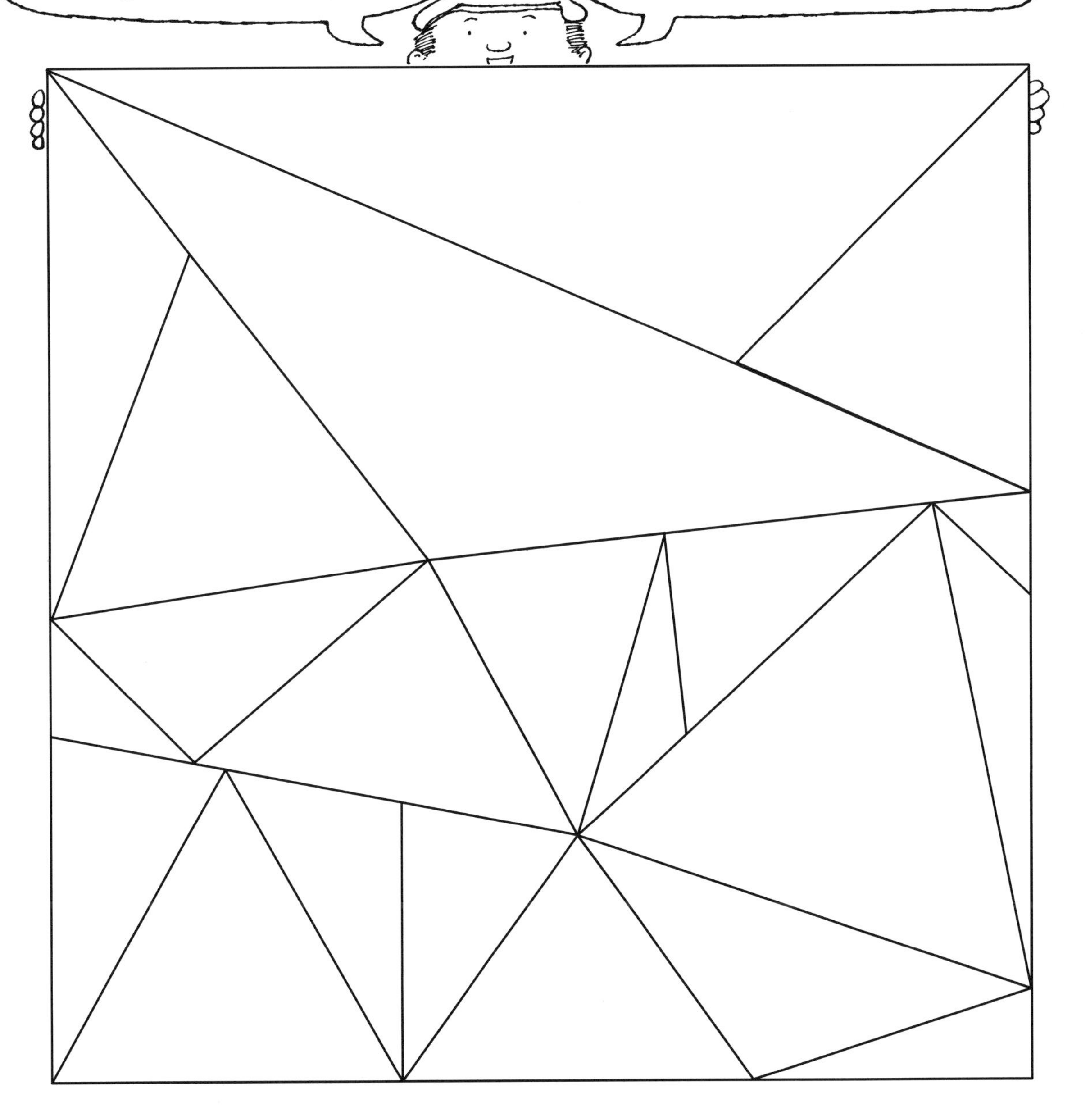

- **Draw a** triangular **pattern of your own.**
- **Record how many of each type of triangle you have drawn.**

Use a ruler.

Teachers' note Before beginning this activity, discuss the properties of isosceles and equilateral triangles. Point out that some of the triangles are neither isosceles nor equilateral and, if appropriate, introduce the term 'scalene', meaning a triangle that has no sides that are equal and no angles that are equal. Some children may need rulers for the main activity.

Tricky sticky shapes

Clare is sorting some sticky paper shapes.

Some are regular and some are irregular.

• Join each shape to the correct envelope.

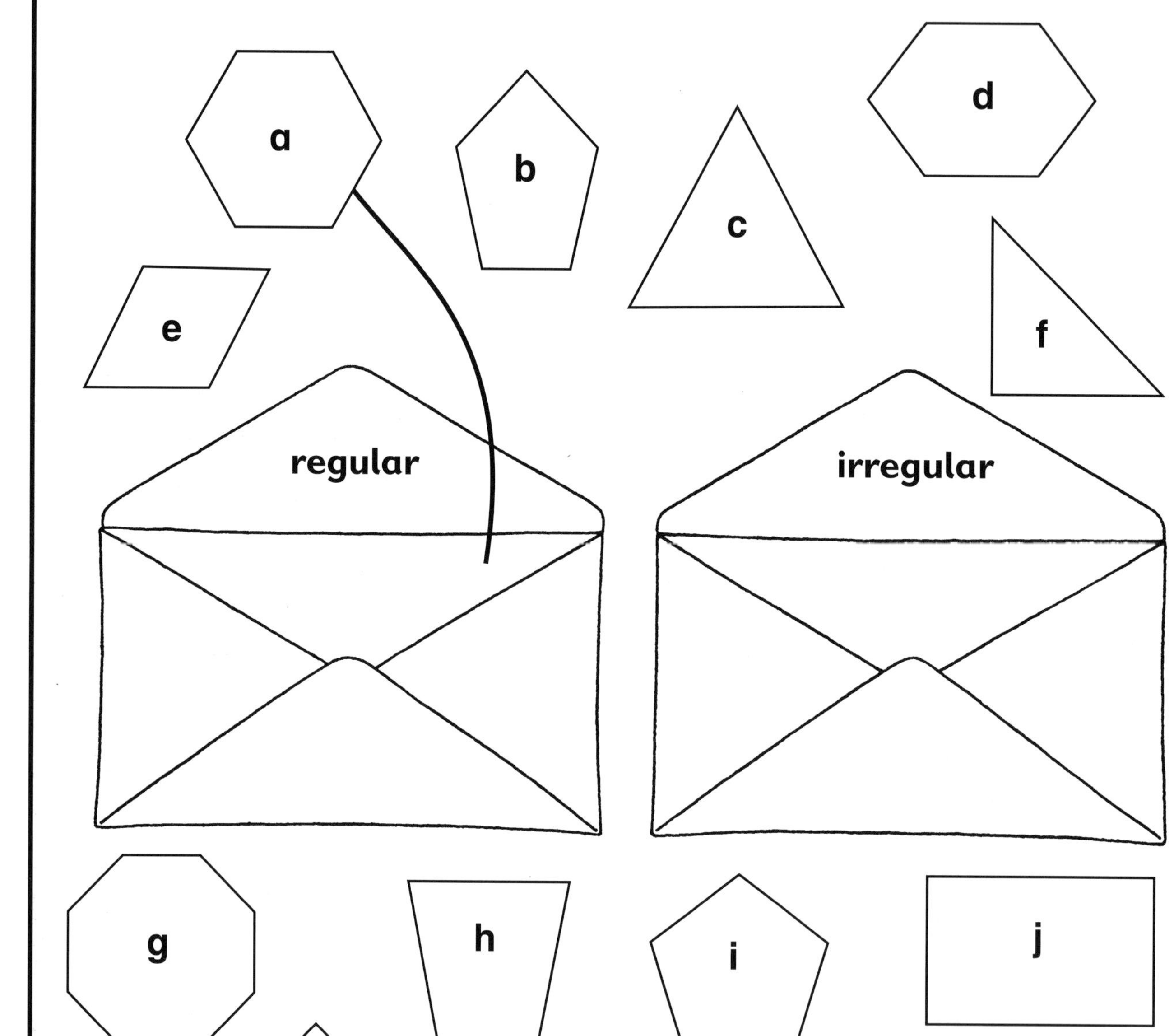

• What is another name for a regular triangle?

____

Teachers' note A regular shape must have all sides equal and all angles equal. Children often think that all rectangles are regular as they have equal angles. They must have equal length sides too, i.e. a square.

Reflection detection

- **Cut out the triangular cards.**
- **Match each shape to its** reflection.

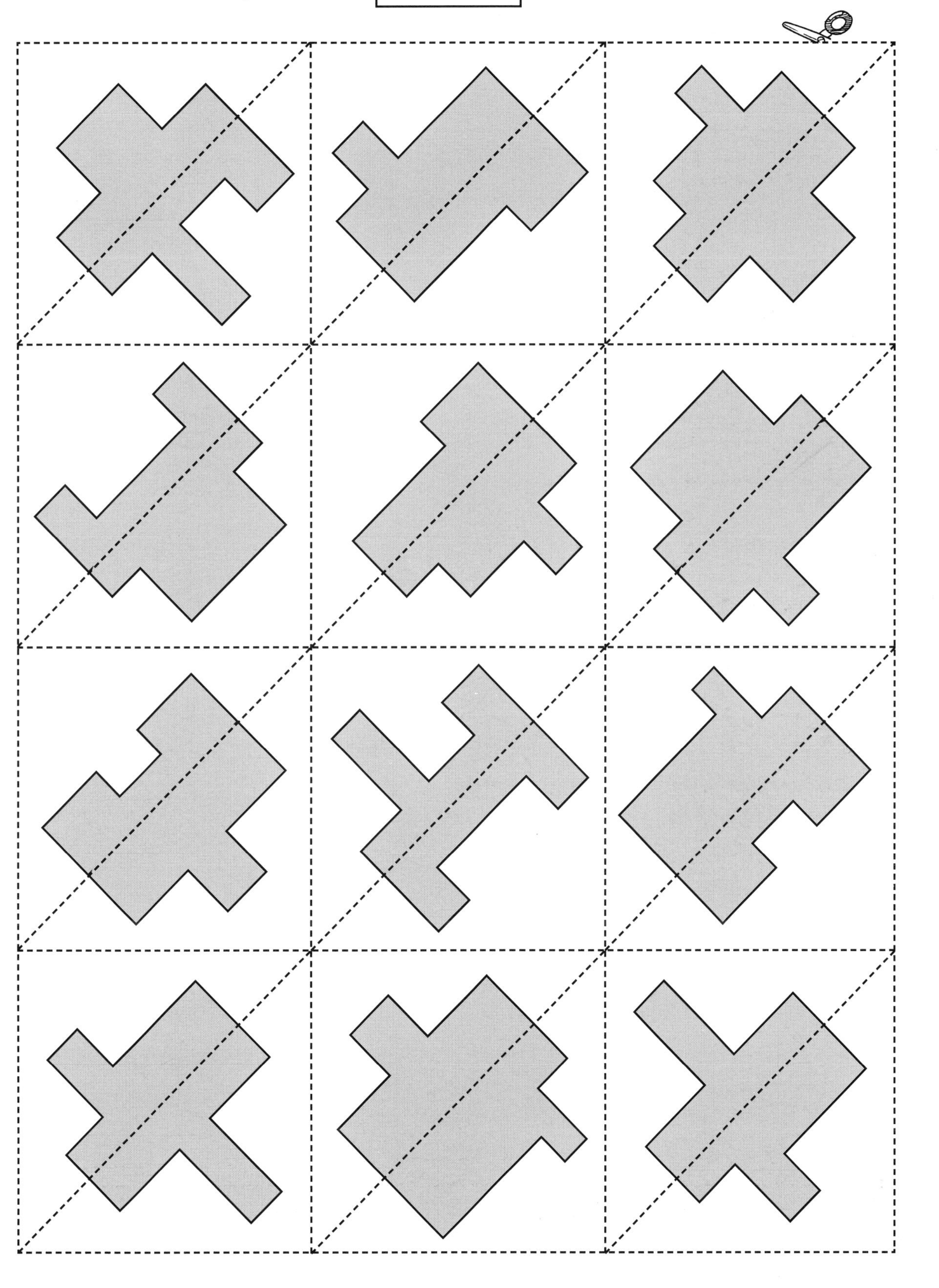

Teachers' note Ensure that the children realise that the diagonal line is the mirror line. The pairs of cards can be glued down, or they can be numbered, for example, 1 and 1, 2 and 2, and so on. Some children may find it easier to hold the mirror lines horizontally when they are matching.

Sketching reflections

- **Sketch the reflection of each shape in the dotted mirror line. One has been done for you.**

Use a mirror and a ruler to help you.

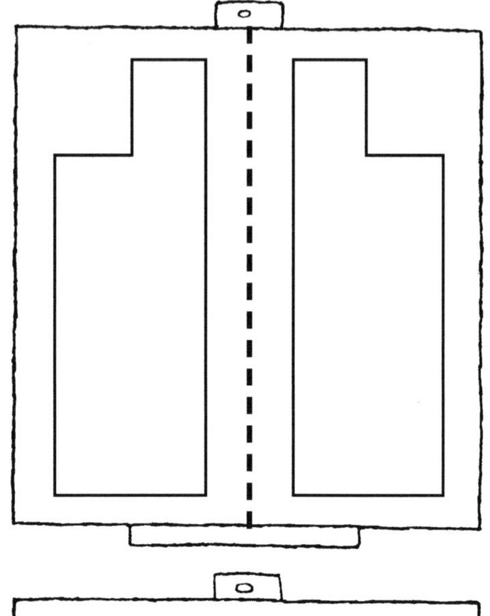

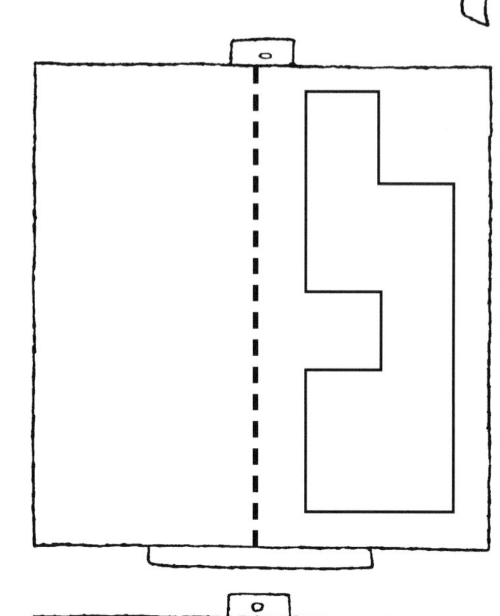

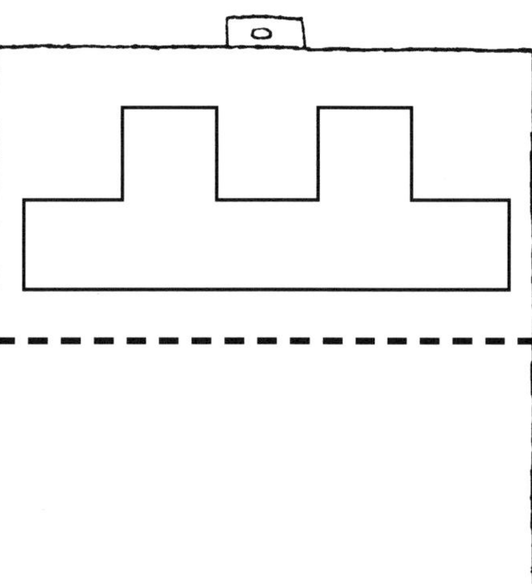

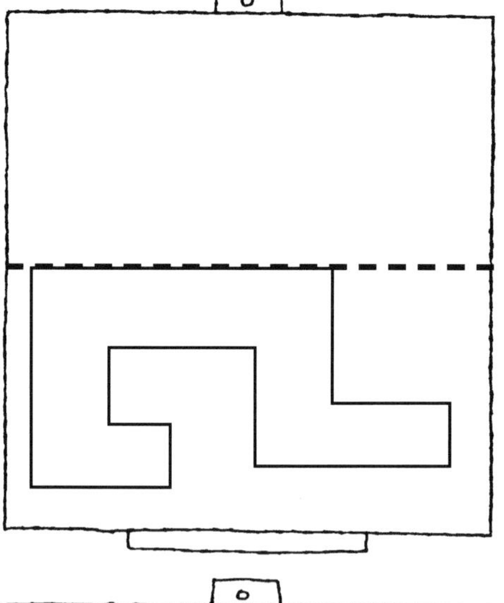

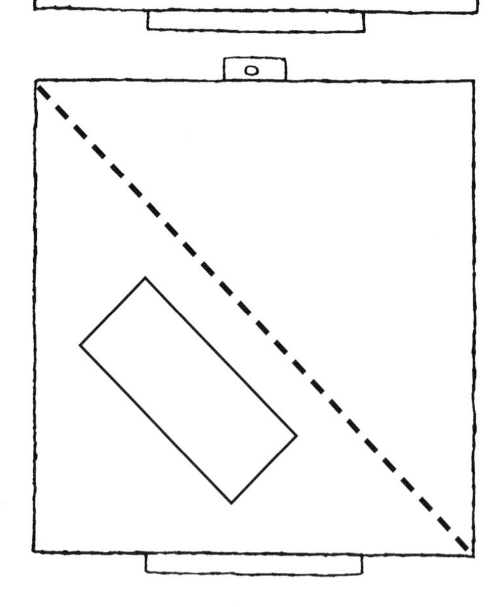

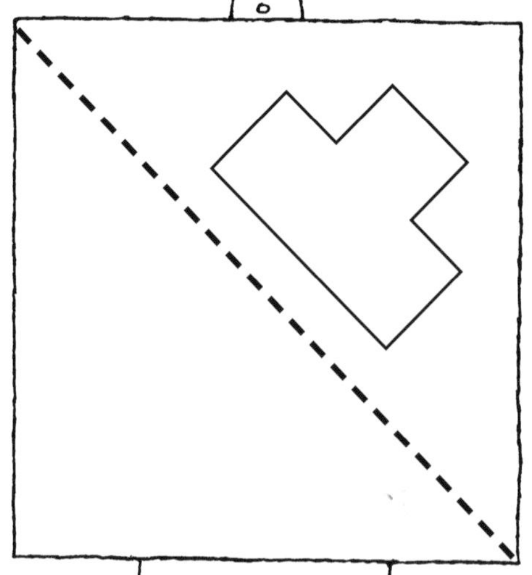

Teachers' note Encourage the children to notice that equivalent points are the same distance from the line of symmetry, and corresponding lengths are the same.

Seeing double

- **Complete the pattern so it has** four lines of symmetry**. Draw the reflections. Colour the pattern.**

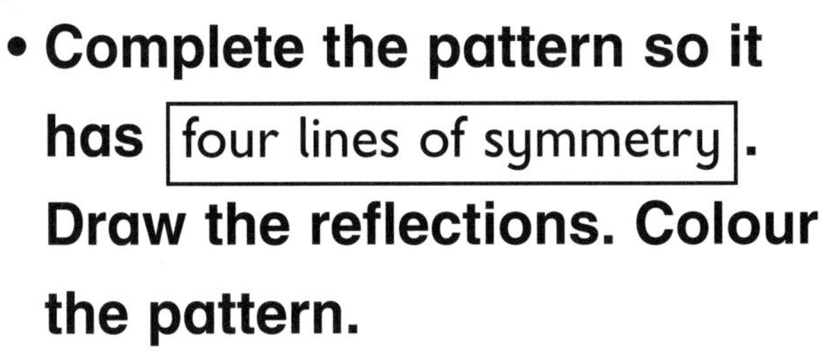
Draw the reflection in the mirror line marked (1), then (2), then (3), and so on.

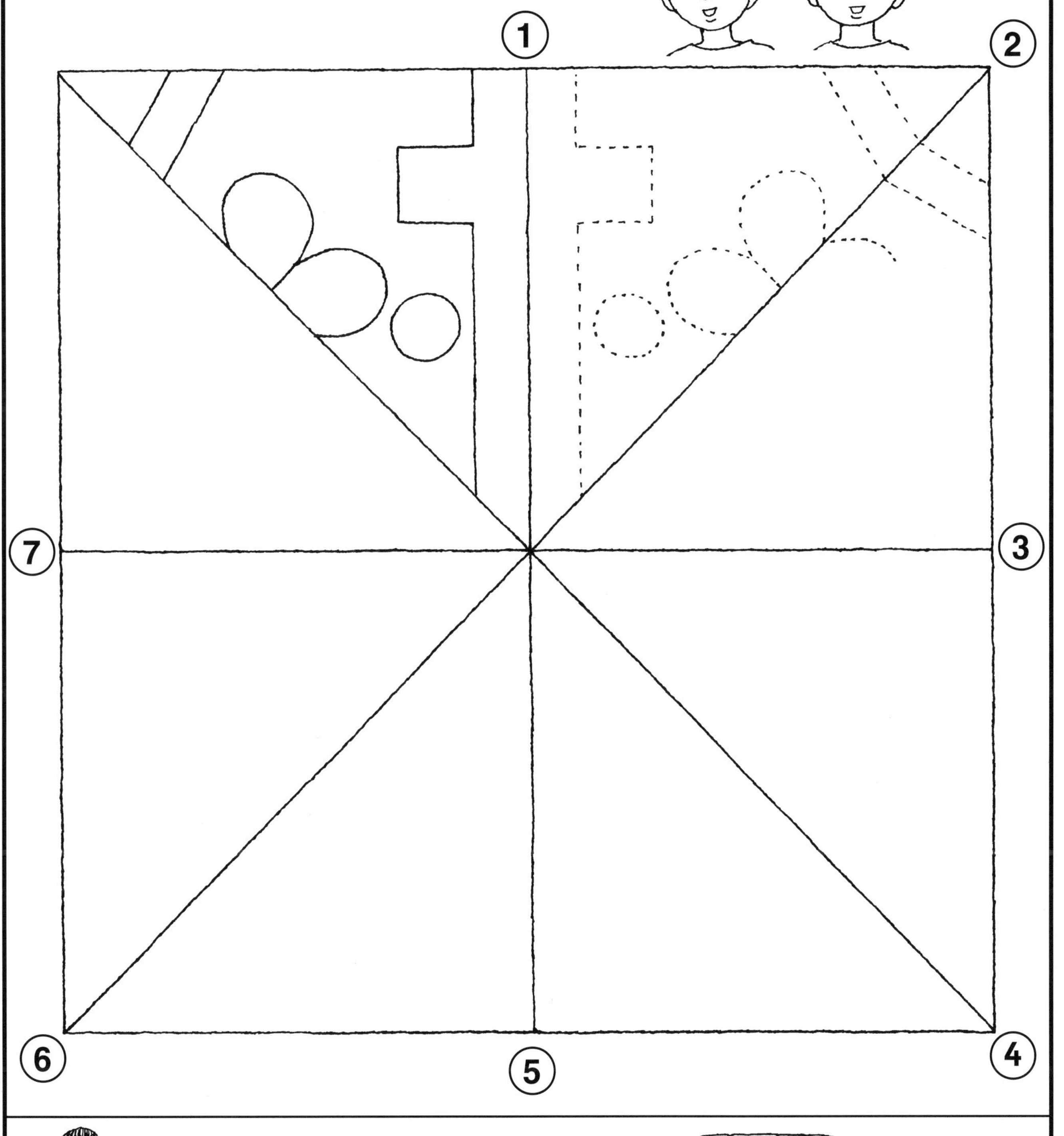
1
2
7
3
6
5
4

Now try this!

- **Draw your own pattern with** four lines of symmetry**.**

Use squared paper.

Teachers' note Ensure that the children realise that a line of symmetry goes right across the shape, for example, line (1) and (5) together are the same line of symmetry. The children, however, will have to build up each part of the reflection using line (1) first and then line (5) later.

How many lines?

- **Tick the chart to show how many** lines of symmetry **each shape has.**
- **Draw the lines of symmetry on the shape.**

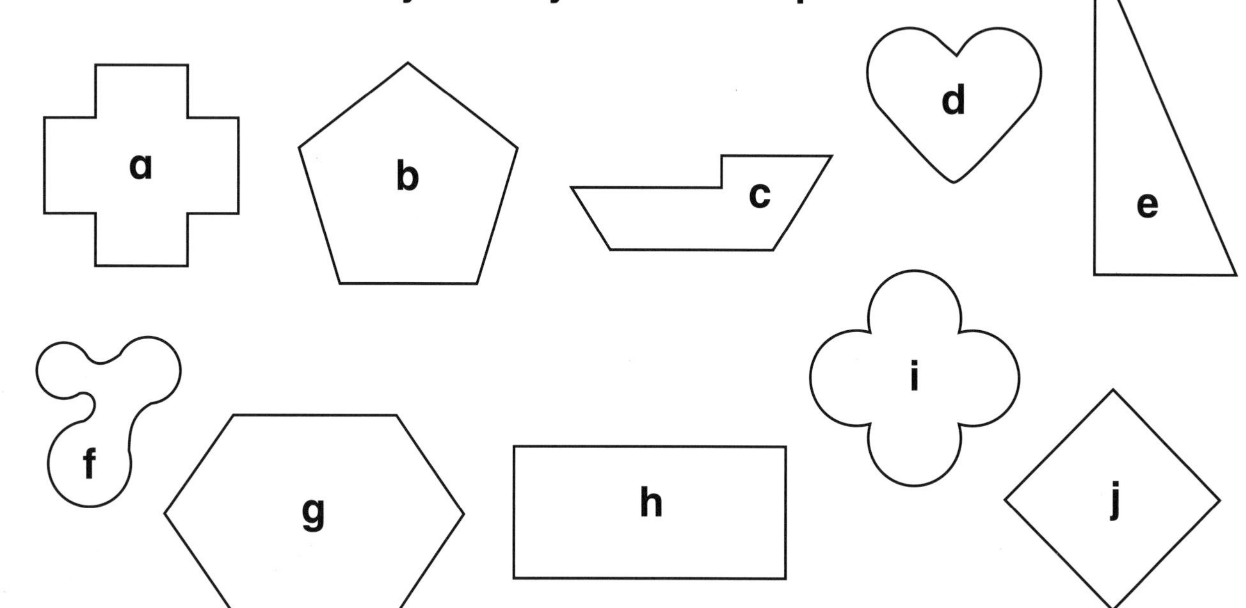

shape	0 lines of symmetry	1 line of symmetry	2 lines of symmetry	more than 2 lines of symmetry
a	✗	✗	✗	✔
b				
c				
d				
e				
f				
g				
h				
i				
j				

- **Draw your own shape to match each column of the chart.**

Teachers' note Children often think that a rectangle has more than two lines of symmetry, mistaking the diagonals for mirror lines. Demonstrate that the diagonals are not mirror lines by folding a sheet of paper. Some children may need a second copy of this sheet to enable them to cut out the shapes and fold them to check.

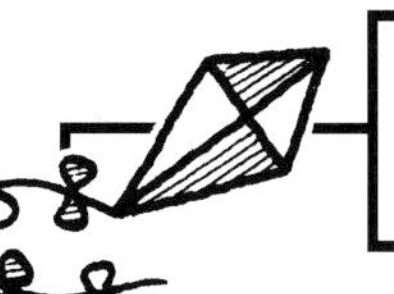

Translating borders

These wallpaper borders show shapes that are translated**.**

- **Continue each pattern.**

When you **translate** a shape, you slide the shape without changing it in any way.

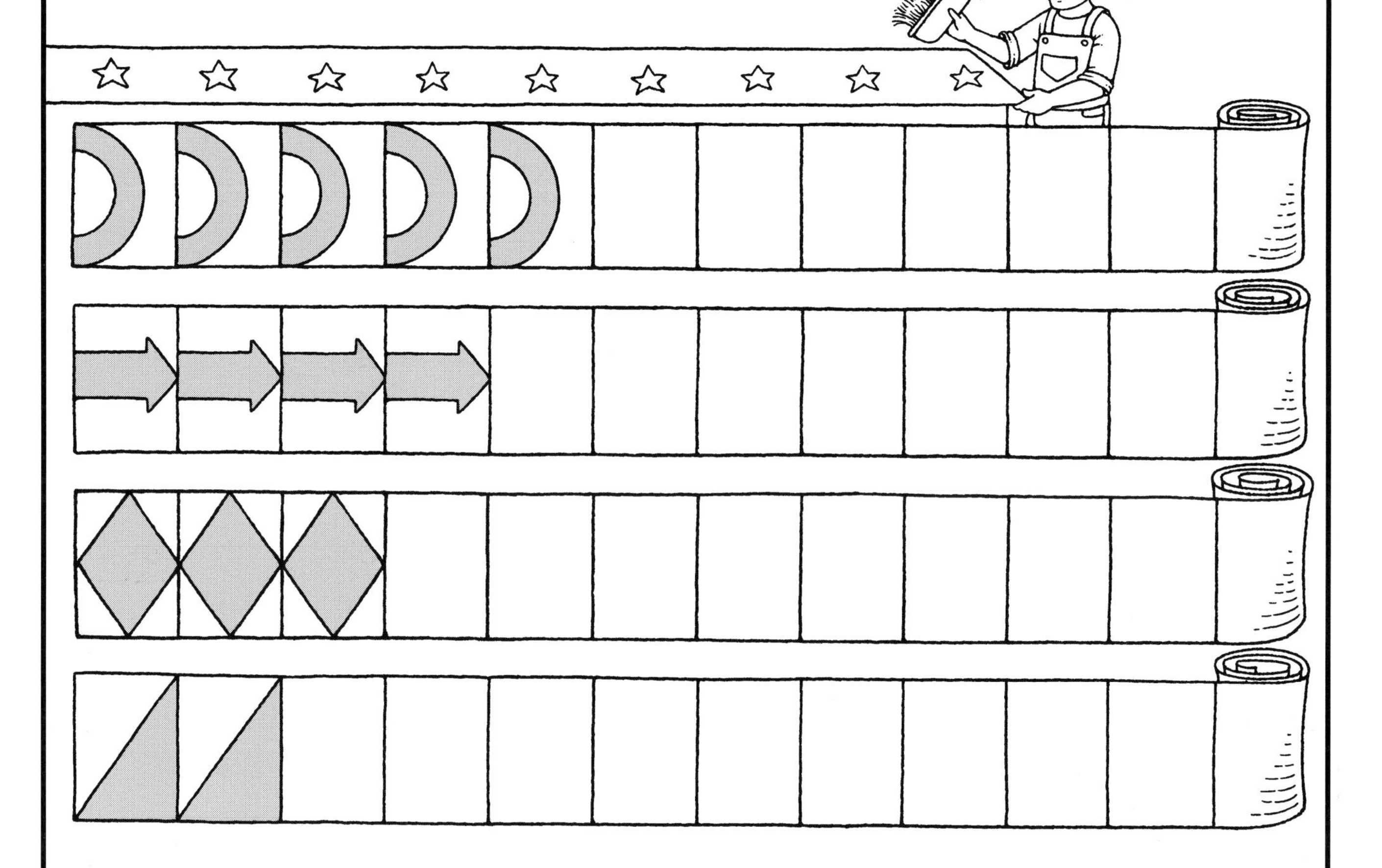

- **Draw two patterns of your own.**

Make sure you translate the shape each time.

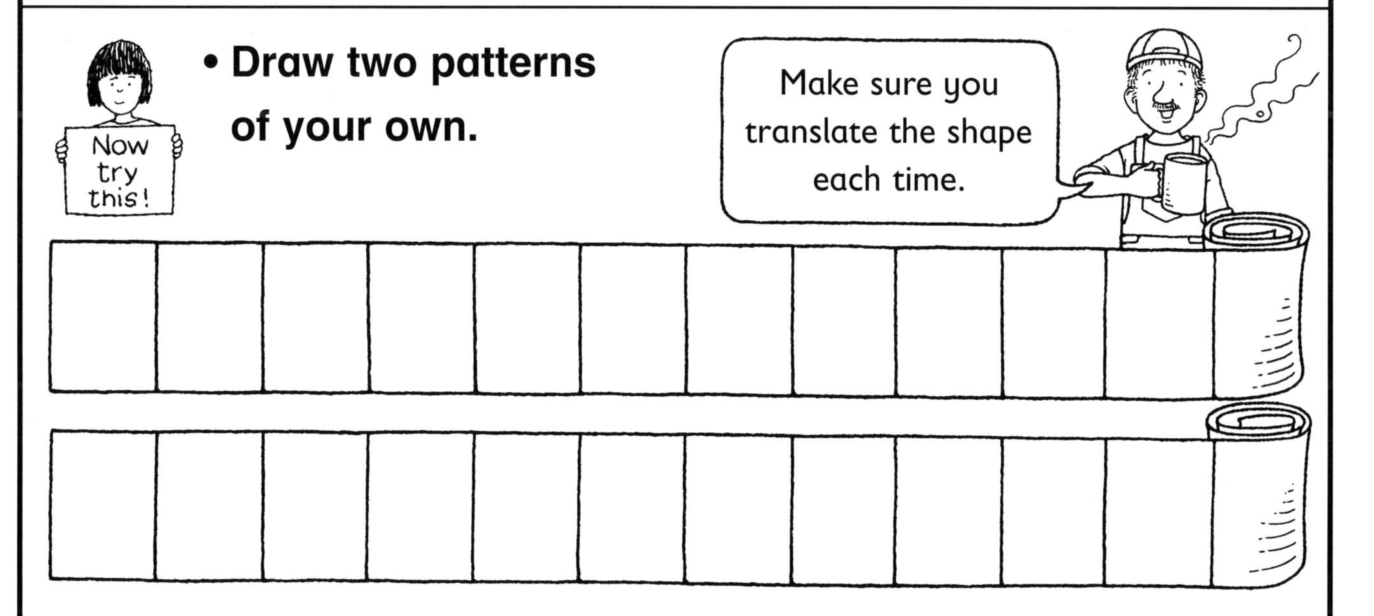

Teachers' note As a further extension, you could ask the children to use long strips of paper and simple cardboard templates or printing blocks to make their own display-sized borders.

Water shortage

- **Write six sets of** co-ordinates.

Use the numbers **0** to **6**.

(,) (,) (,) (,) (,) (,)

- **Mark your co-ordinates on the desert map with water bottles. These are your water supplies.**

Key

water supply

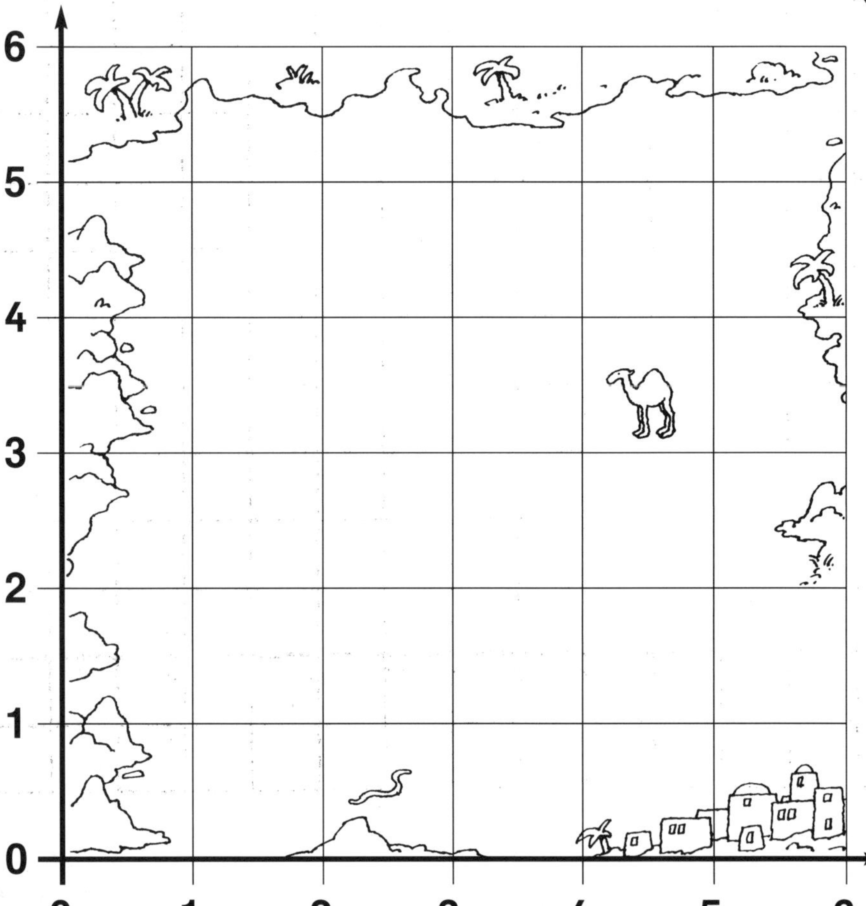

Your opponent's water supplies

(,)

(,)

(,)

(,)

(,)

(,)

- **Play this game with a partner.**

☆ Take turns to call out co-ordinates. Your opponent says 'hit' if you have found one of his/her bottles, or 'miss' if you haven't.

☆ Write the co-ordinates of your opponent's bottles as you find them.

☆ The winner is the first player to find all their opponent's water supplies.

Teachers' note Revise how to write co-ordinates and remind the children that the first co-ordinate shows how many across to move from the origin. Encourage the children to keep a record of all co-ordinates called out to avoid repetition, and so that they can check at the end for errors.

Making squares

- **Play this game with a partner.**

☆ Take turns to roll two dice. Write the numbers as a co-ordinate.

Example: (2, 5) or (5, 2)

☆ Mark the co-ordinate on the grid with a cross (if it is not already marked).

☆ The winner is the first player to join four crosses in a **square**.

You need: two dice.

The square can be any size or orientation.

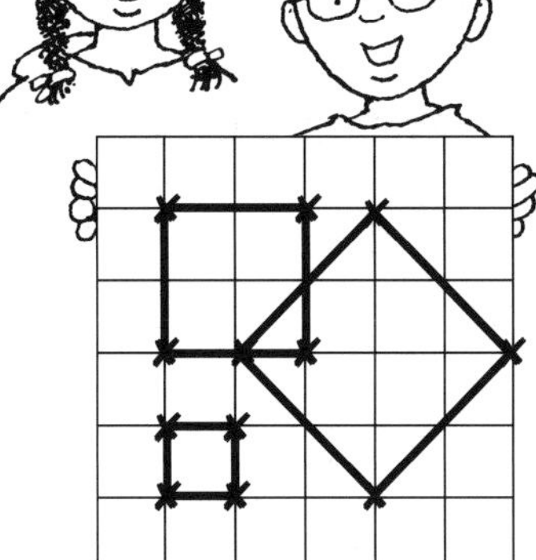

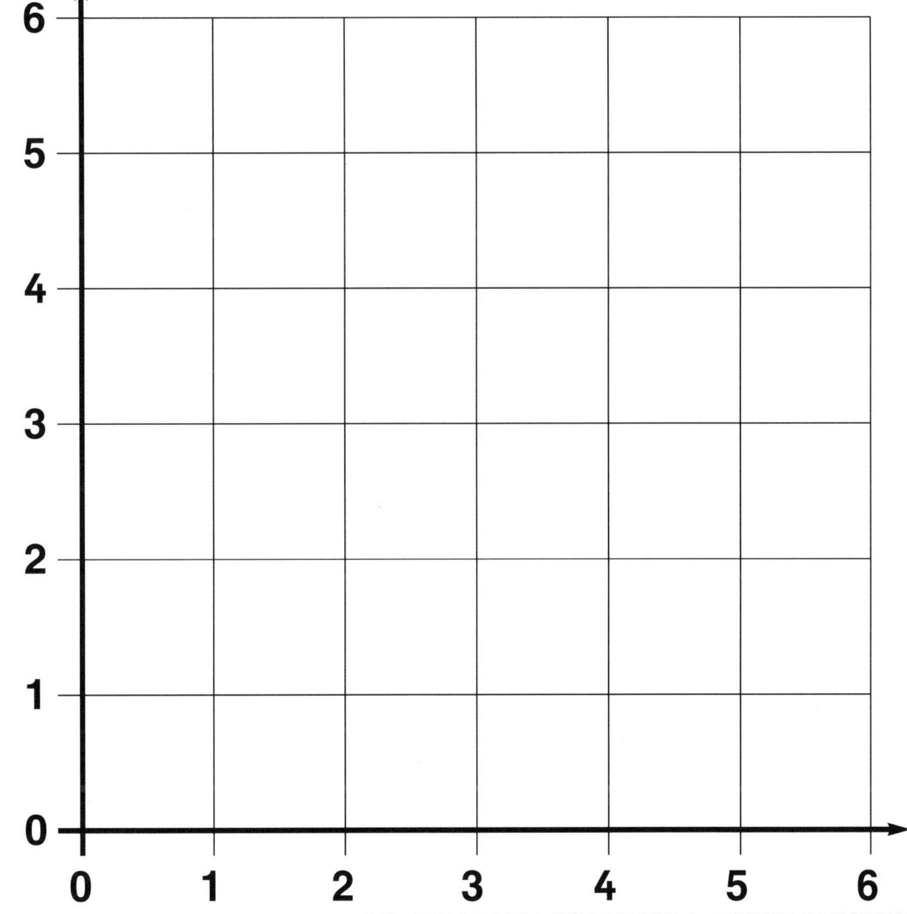

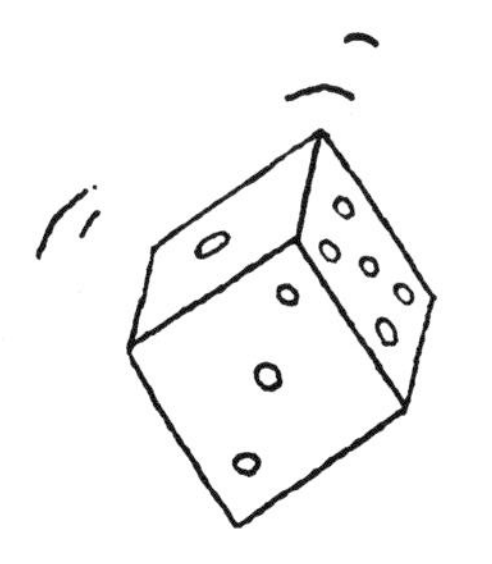

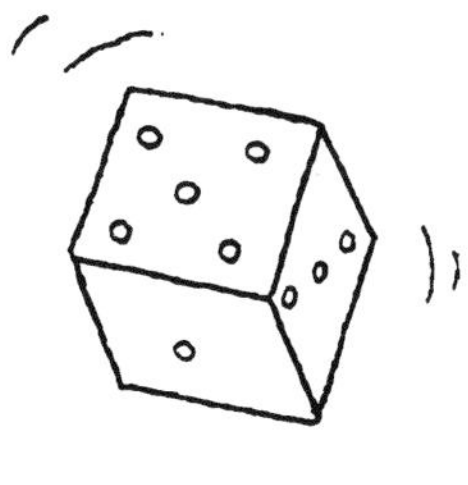

- **Write your co-ordinates here.**

Name:	Name:

Teachers' note Each pair of children needs one copy of the sheet. They should use different colours to mark their co-ordinates. You could use phrases like, 'along the corridor and up the stairs' to remind the children which way to travel from the origin. If necessary, explain 'orientation'.

Sweet search

Lotty has hidden some sweets from her friends!

- **Plot the co-ordinates and join the points to find out where they are.**

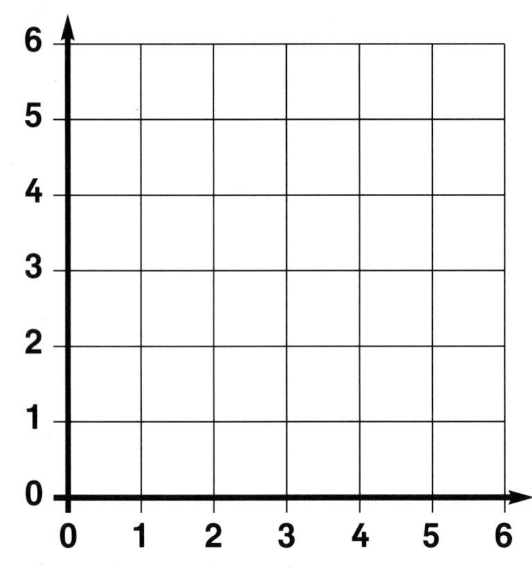

Draw a line from:
(1, 1) to (1, 5)
(2, 1) to (2, 5)
(2, 5) to (5, 1)
(5, 1) to (5, 5)

Draw a line from:
(4, 3) to (6, 3)
(9, 1) to (7, 1)
(2, 5) to (2, 1)
(7, 3) to (8, 3)
(7, 1) to (7, 5)
(1, 5) to (3, 5)
(6, 1) to (6, 5)
(7, 5) to (9, 5)
(4, 1) to (4, 5)

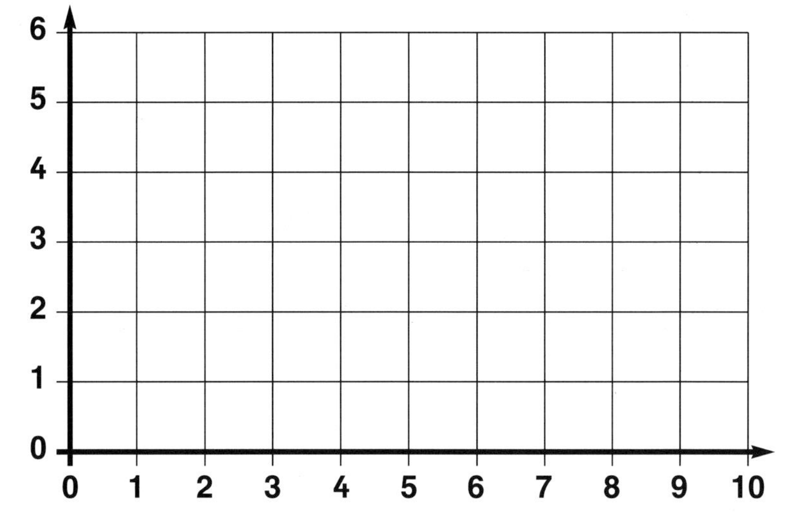

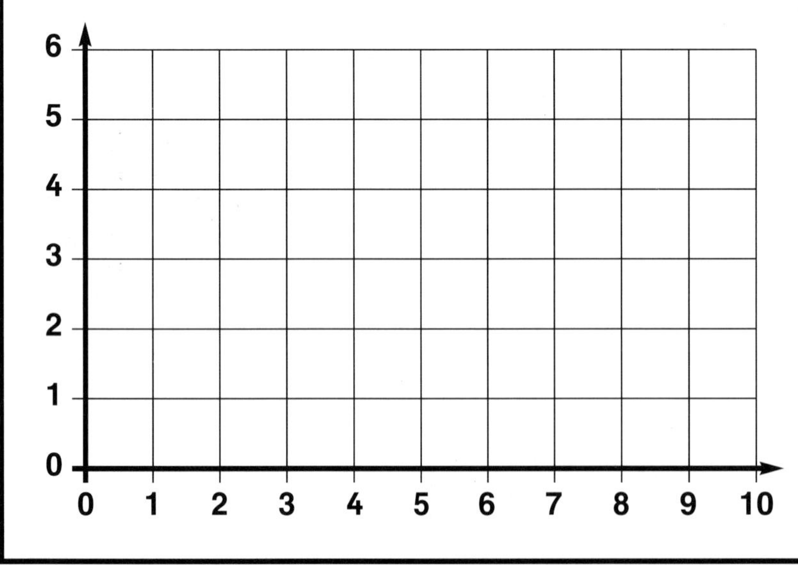

Draw a line from:
(4, 5) to (4, 1)
(5, 5) to (8, 1)
(1, 5) to (3, 5)
(8, 1) to (8, 5)
(2, 1) to (2, 5)
(5, 1) to (5, 5)

Teachers' note Remind the children that the first co-ordinate shows how many across to move from the origin. To avoid confusion, encourage the children to cross off each instruction as the line is drawn. As an extension, the children could write the co-ordinates for a short secret message to give to a partner.

Horizontal and vertical

- **Write whether these are** horizontal, vertical or neither.

1. lines on a musical stave

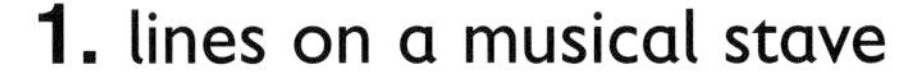

horizontal

2. fence posts

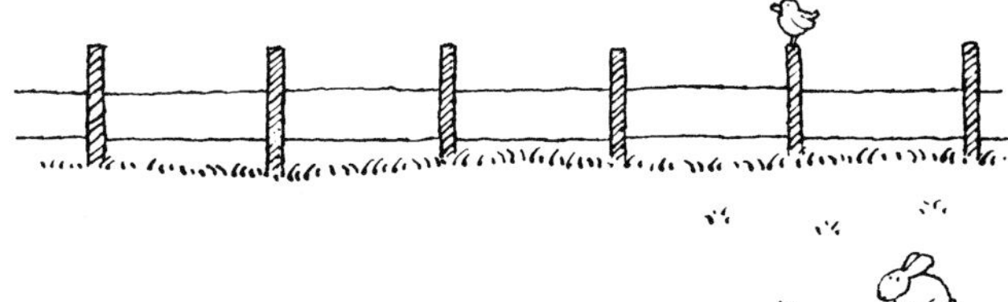

____

3. sides of this lighthouse

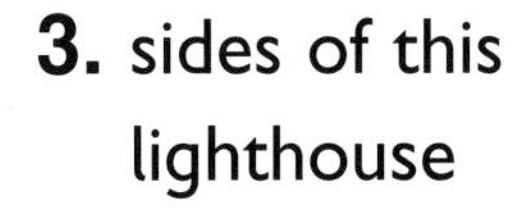

____

4. horizon over the sea

____

5. flash of lightning

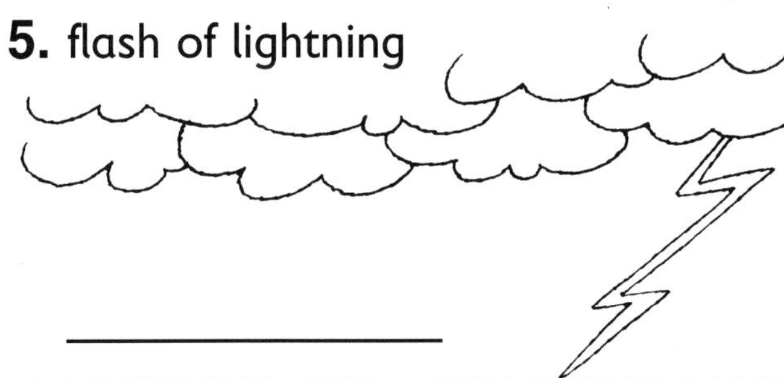

____

6. flagpole

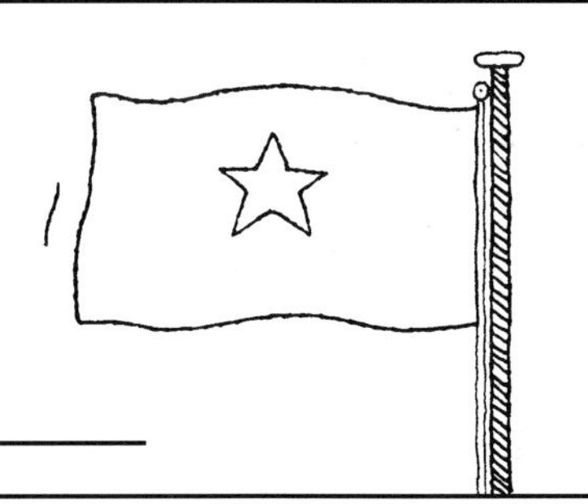

____

7. rungs on a ladder

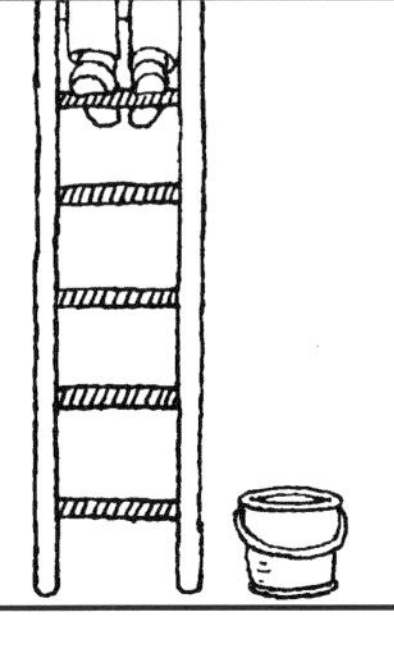

____

8. this washing line

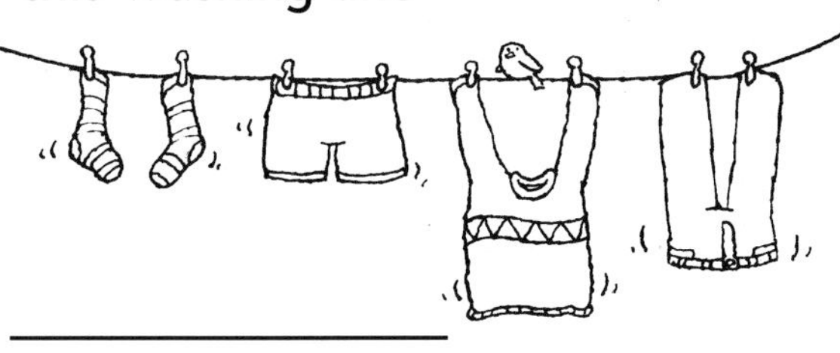

____

9. lamppost

____

10. these garden canes

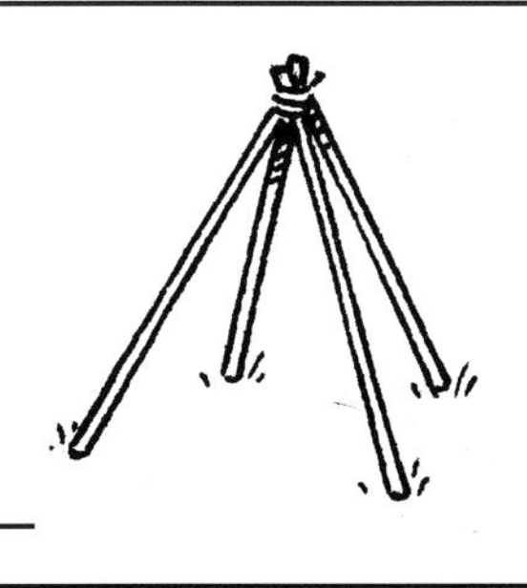

____

- **Write two more things that are:**

horizontal | vertical | neither

Teachers' note Revise the terms 'horizontal' and 'vertical'. The children can also be introduced to 'parallel' and 'perpendicular' as part of this activity. Ensure that they appreciate that horizontal and vertical lines must be straight.

Blast off!

You and your partner are racing to leave the space station. You cannot blast off until you have 30 litres **of fuel.**

- ☆ Place your counter on 'start'. Take turns to roll the dice.
- ☆ Move **one** dot in the direction shown in the key. If you cannot move, stay where you are. Only one counter can be on a dot.
- ☆ When you reach the fuel, write it down. Jump back to 'start'.
- ☆ Continue until you can blast off!

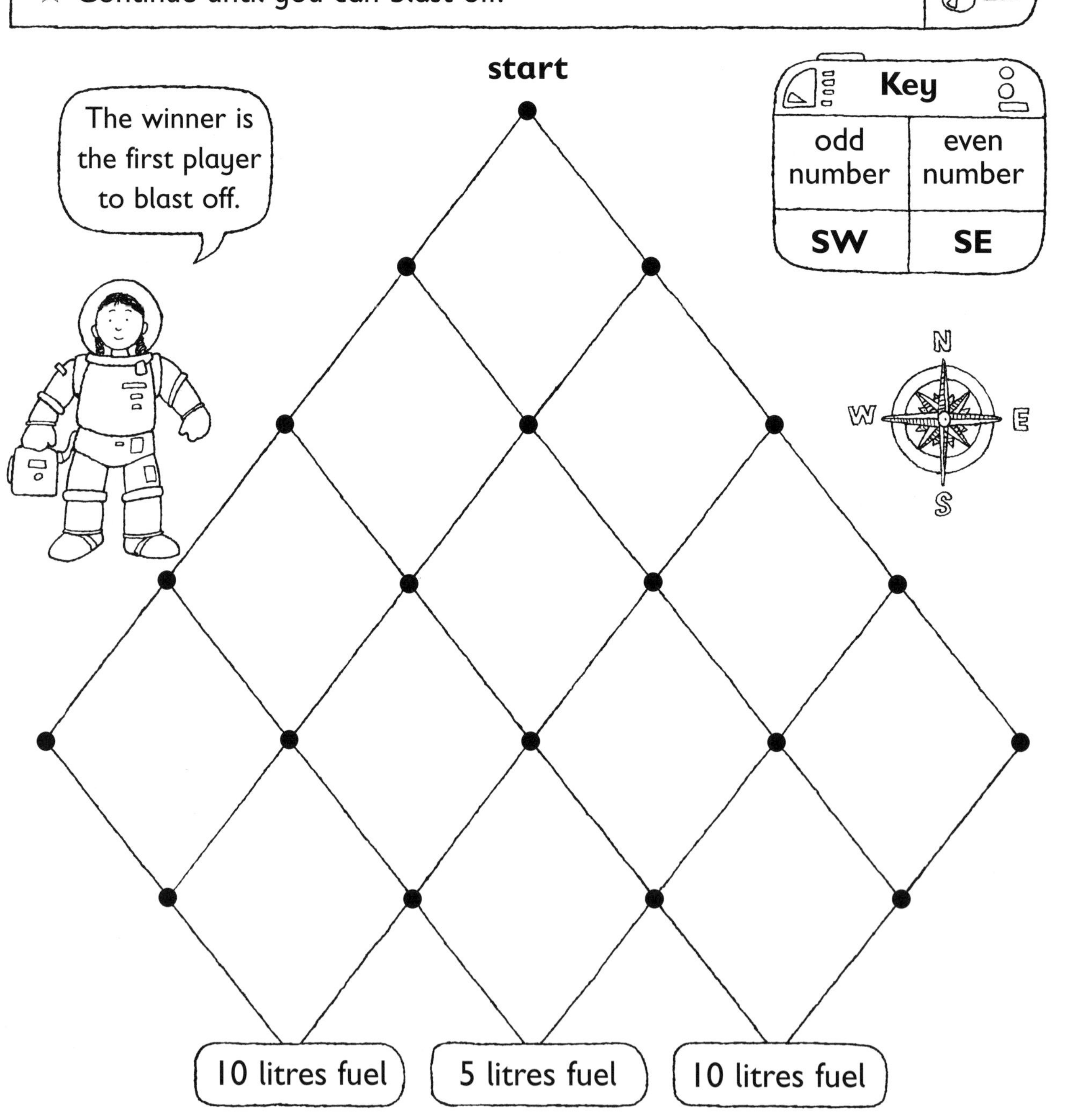

Teachers' note Introduce NE, NW, SE and SW. Ensure the children realise that once they have collected fuel they should return to 'start' and continue until they have 30 litres. Remind them to keep a record of the number of litres collected. As an extension, the children could record all the possible ways of collecting 5 litres, for example: SW, SW, SE, SE, SW, SE.

Cops and robbers

• **Play this game with a partner.**

You need: 2 counters.

☆ Cut out the cards. Spread them out face down.

☆ Place your counter on 'cop' or 'robber'.

☆ Take turns to reveal a card. Move one block in the direction shown. Replace the card. If you cannot move, stay where you are.

☆ The game ends when the cop catches the robber.

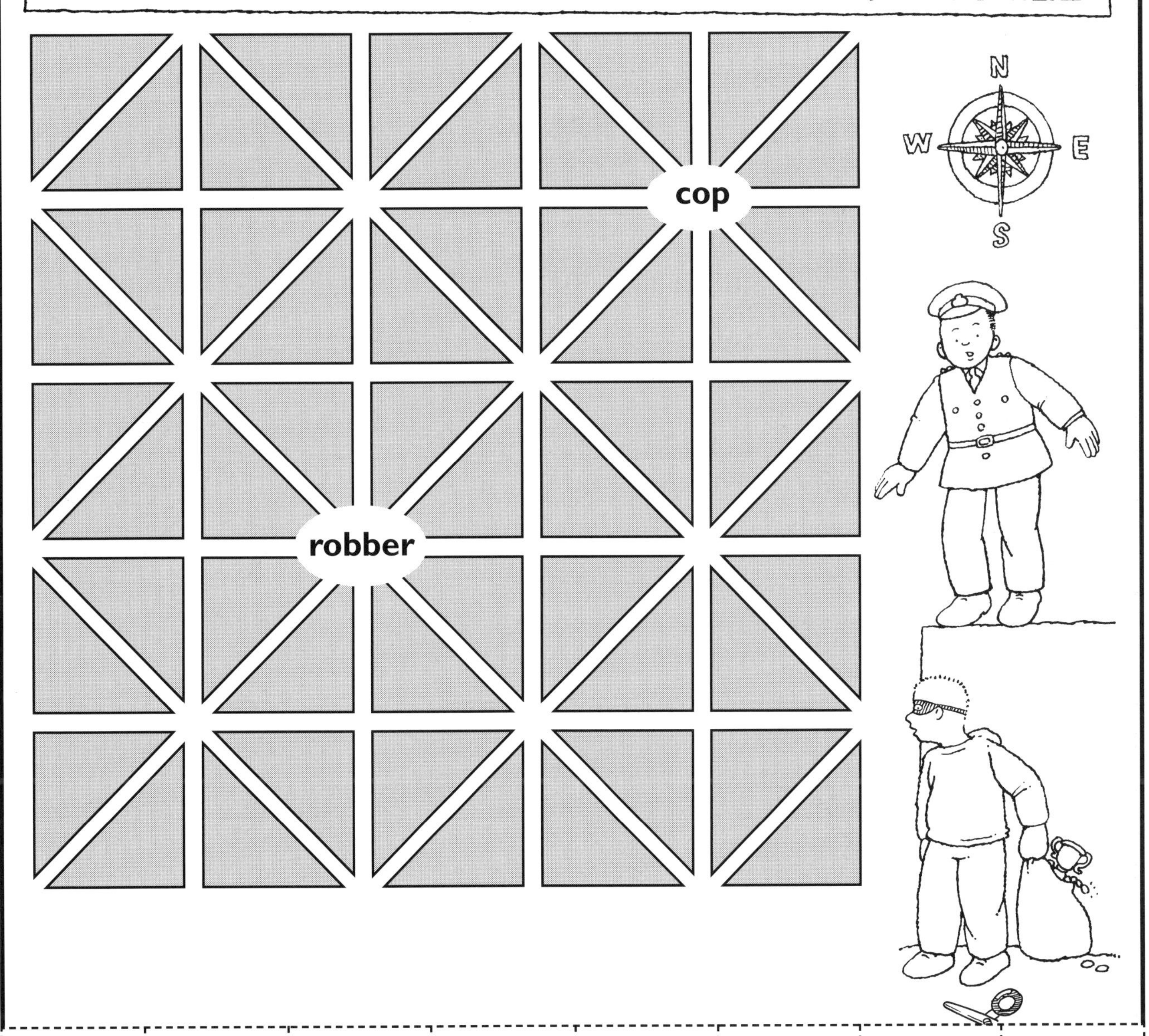

N	S	E	W	NE	NW	SE	SW

Teachers' note Each pair of children needs one copy of the sheet (although if you wish, you could make extra copies to provide more cards). Before beginning the game, explain the term 'block'. At the end of the game, the children could reverse roles and play again. The game can be adapted for use with more than two players by including more cops or more robbers.

Map reading

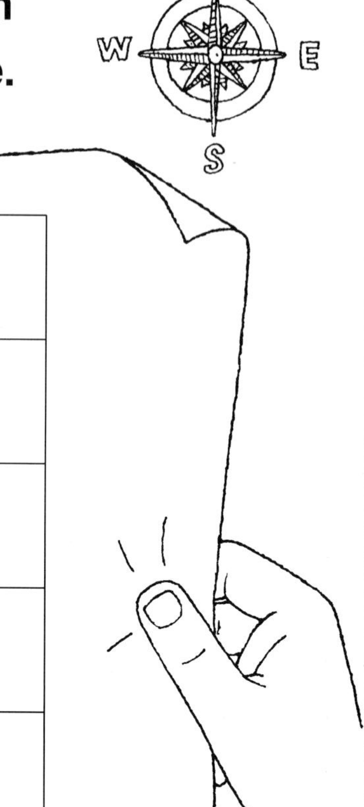

• **Look at the map. Write the compass direction you would move in to get from place to place.**

1. house to church NW
2. bridge to house ________
3. pond to house ________
4. bridge to pond ________
5. bridge to church ________
6. tree to phone box ________
7. pond to church ________
8. phone box to church ________
9. house to tree ________
10. pond to phone box ________
11. phone box to bridge ________
12. church to pond ________

• **Write the directions for going from the tree to the bridge. You must visit all the other places on the way.**

Teachers' note As a further extension, the children could draw their own map and write their own directions.

Snap angle

- Cut out the cards.
- Play 'Snap' with a partner.

360°	a half turn	45°	one right angle
half a right angle	one whole turn	two half turns	four right angles
two right angles	two quarter turns	90 degrees	180°
90°	180 degrees	four quarter turns	a quarter turn
45 degrees	two 90° turns	360 degrees	half of 90°

Teachers' note The children can play matching games, such as Pelmanism, or can sort the cards into groups. The cards can also be used in conjunction with a spinner or arrow, where children take turns to pick a card and move the arrow.

Lucky wheel

In a TV game show, contestants spin a wheel to win a prize.
The wheel returns to 'start' after each turn.

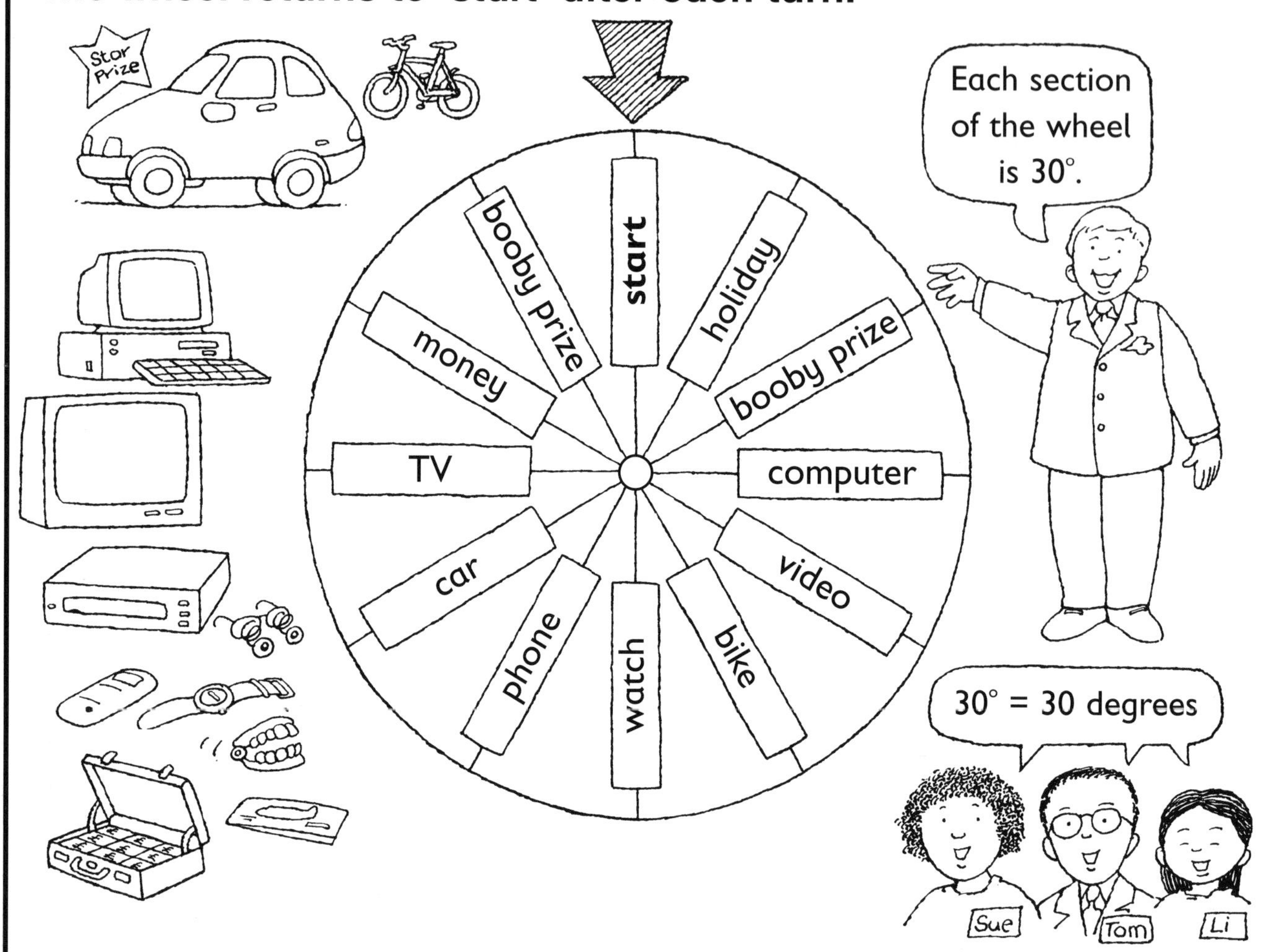

Each section of the wheel is 30°.

30° = 30 degrees

Sue Tom Li

• Write the prize they win if the wheel turns:

1. 30° clockwise

booby prize ____________________

2. 30° anticlockwise

____

3. 90° clockwise

____

4. 90° anticlockwise

____

5. 120° clockwise

____

6. 180° anticlockwise

____

7. 150° clockwise

____

8. 150° anticlockwise

____

Teachers' note Encourage the children to turn the page if necessary through the angles shown. Alternatively, they could cut out the wheel and turn it. As an extension, they could write the turns required to win each item.

Angle art

- Cut out the set squares from the bottom of the page.
- Use them to measure the angles on the picture. Label them.

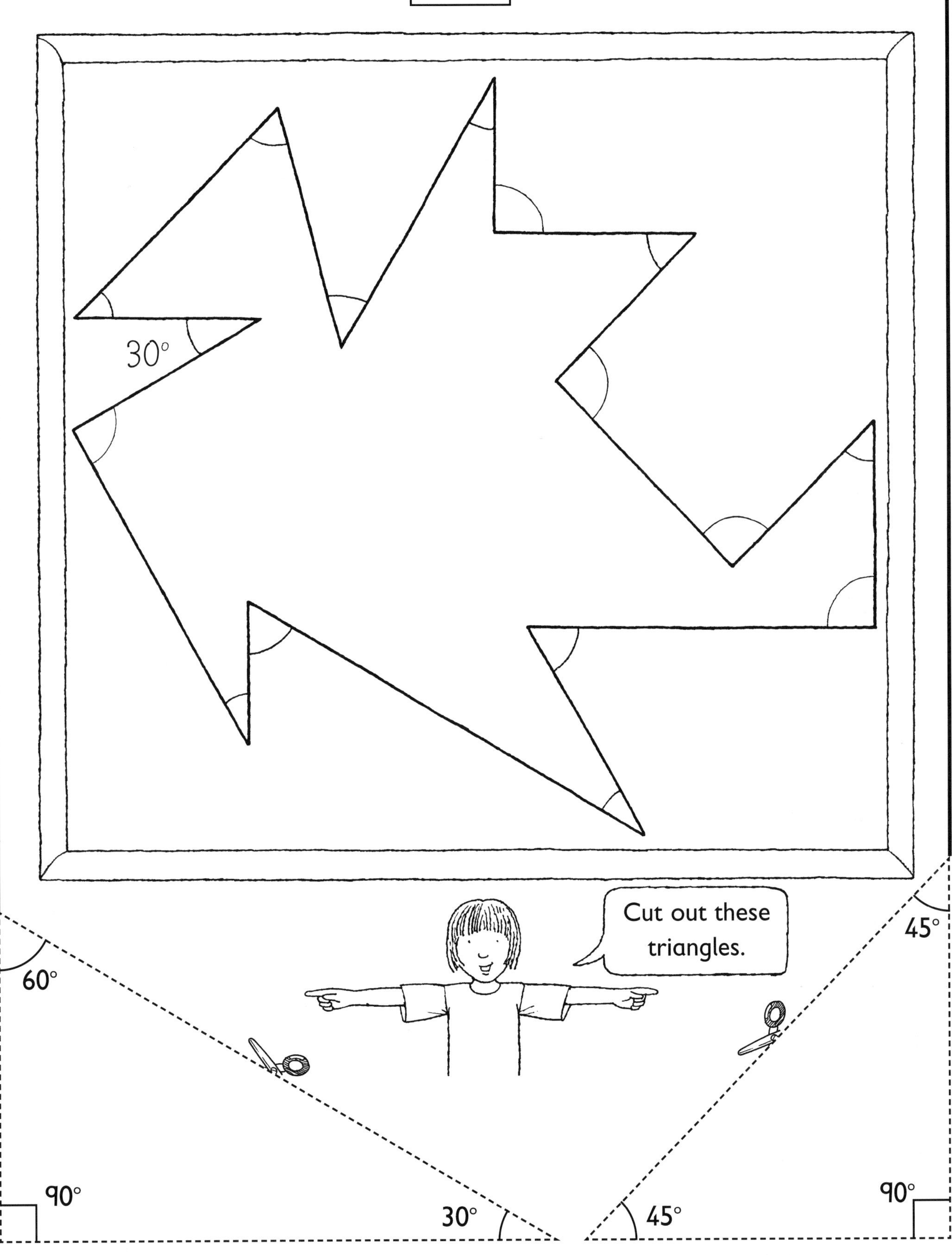

Teachers' note Demonstrate how to position a set square to compare the angles and determine their size. Encourage the children to estimate each angle first. As an extension, the children could make their own angle art pictures.

In a spin

Here is a dial on a washing machine. The arrow turns clockwise as the washing machine cycle moves on.

- **Write how many degrees the arrow turns.**

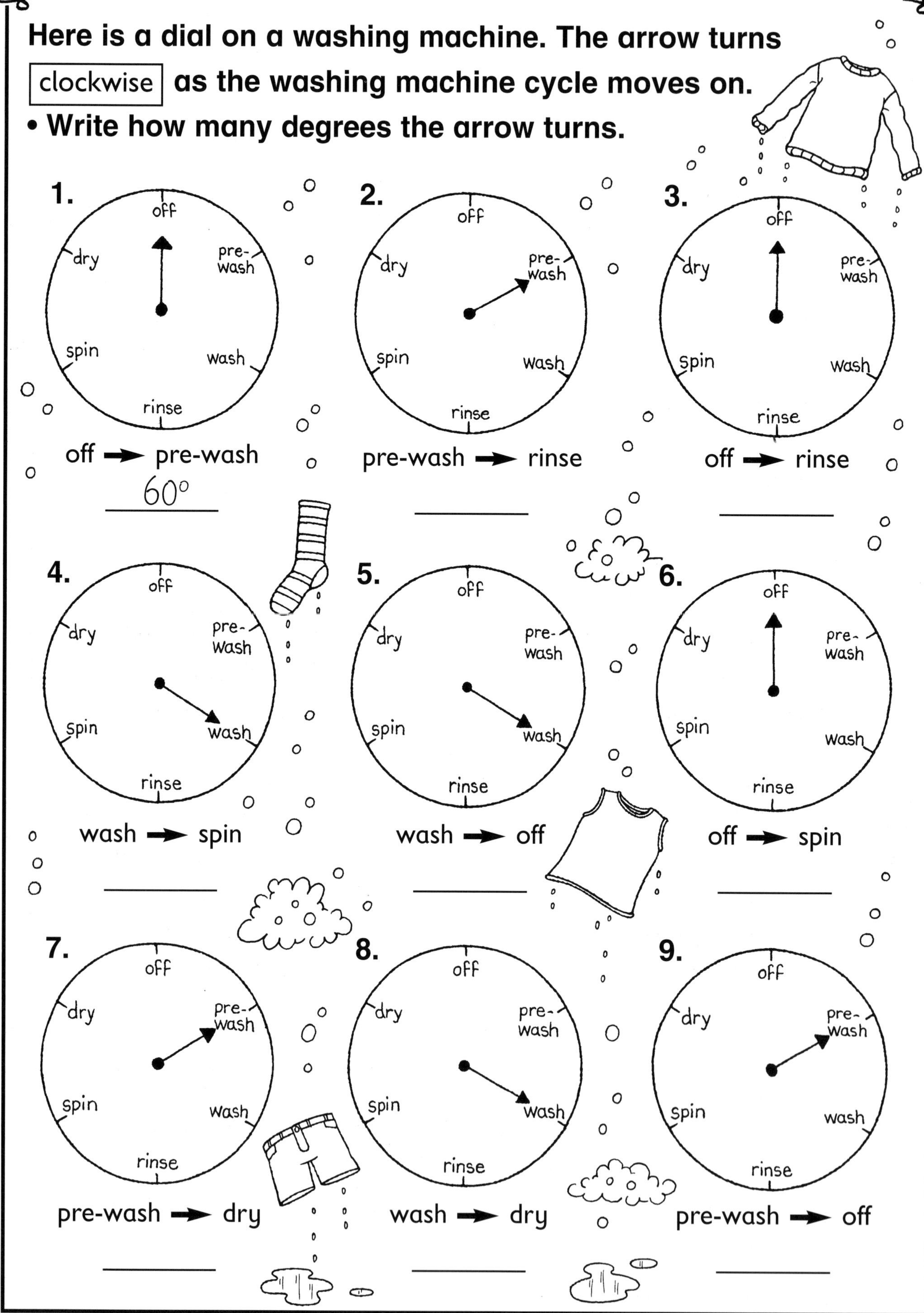

Teachers' note Remind the children that the arrow always turns clockwise and explain that the dial is divided up equally so that each section is 60°. The children could make their own dials to help them with the activity.

Spell it out

- **Look at the spelling wheel.**
- **Follow the instructions to discover the words.**

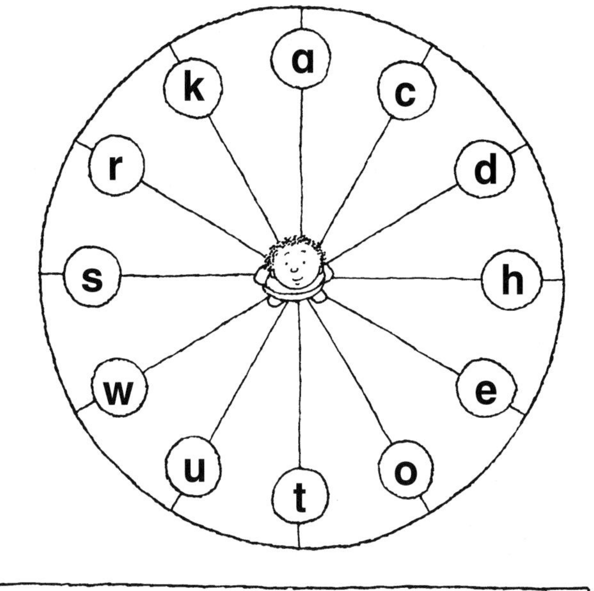

1. Face **c**. Turn 30° anti-clockwise.
Turn 180° clockwise.

cat ______

2. Face **r**. Turn 60° clockwise.
Turn 180° clockwise.

______

3. Face **c**. Turn 120° clockwise.
Turn 90° clockwise.

______

4. Face **h**. Turn 90° anti-clockwise.
Turn 60° anti-clockwise.
Turn 180° clockwise.

______

5. Face **d**. Turn 150° clockwise.
Turn 180° clockwise.
Turn 60° anti-clockwise.

______

6. Face **s**. Turn 180° clockwise.
Turn 90° anti-clockwise.
Turn 60° anti-clockwise.
Turn 30° clockwise.

______

7. Face **h**. Turn 60° clockwise.
Turn 150° clockwise.
Turn 30° anti-clockwise.
Turn 150° anti-clockwise.

______

- **Follow these instructions. What do they spell?**

Face C. Turn 90° anticlockwise. Turn 180° clockwise.
Turn 120° anticlockwise. Turn 180° clockwise. Turn 30° clockwise.
Turn 90° clockwise. Turn 180° clockwise. Turn 150° clockwise.

______

Teachers' note Explain that each section of the wheel is 30°. Encourage the children to notice that a 180° turn clockwise or anticlockwise gives the same result. As an extension, the children can make up their own words and write directions for them. A clock face with a moveable hand and the numbers covered by the 12 letters could be used for children who need more help.

The larger angle

- **Cut out the cards.**
- **Pick pairs of cards. Compare the angles.**

Example: Angle **a** is larger than angle **f**.

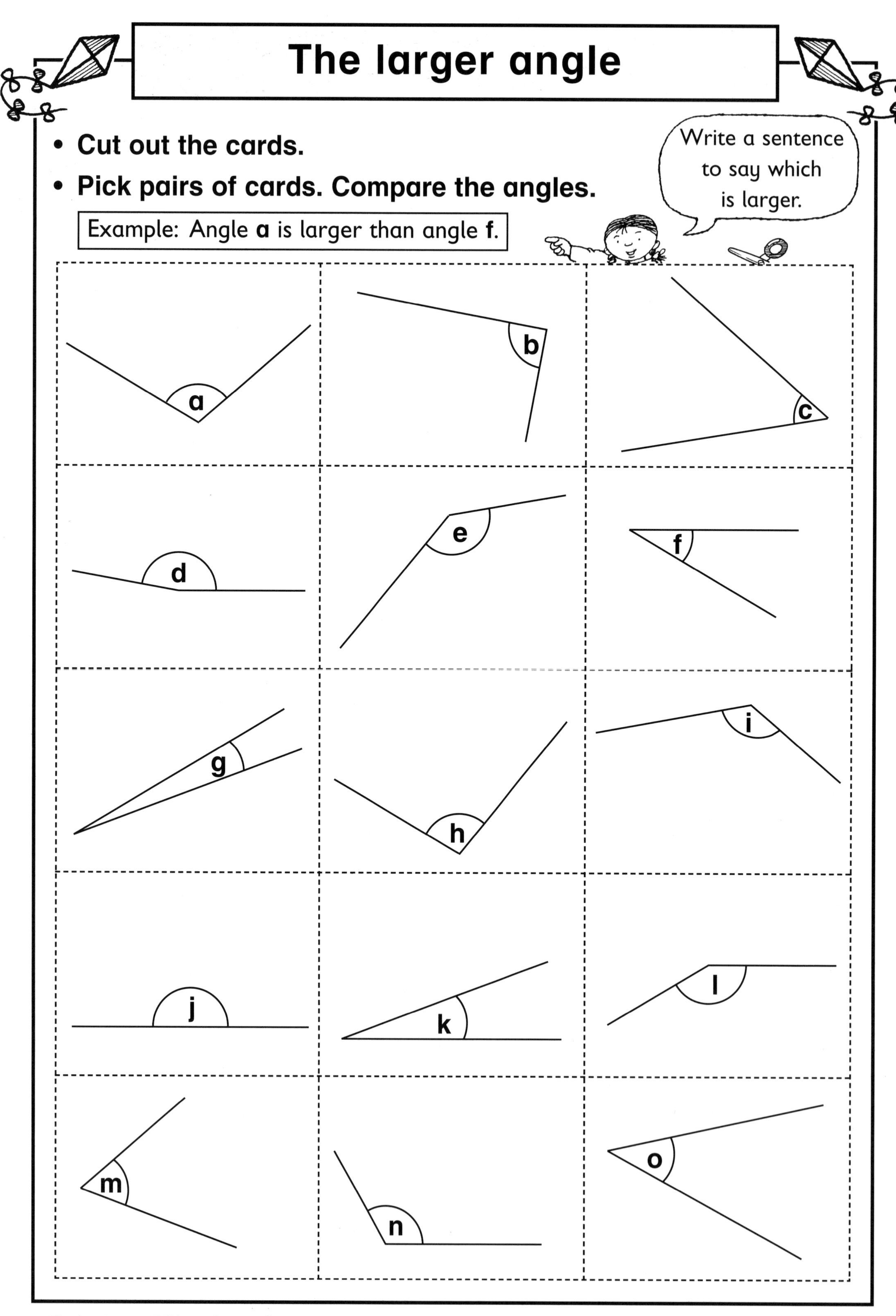

Teachers' note Children often confuse angle with lengths of lines, distances between end points, areas within arcs, and so on, rather than appreciating it as an amount of turn. Some children may need to trace the angles to help them make comparisons. As an extension, the children could arrange the cards in order.

Answers

p 6

Now try this!

$\frac{1}{10}$ m, $\frac{1}{4}$ m, $\frac{1}{2}$ m, $\frac{3}{4}$ m, 1·2 m, 1·4 m, 2 m, 2·1 m,
3·4 m, 4 m, 4·1 m, 5 m, 8 m

p 7

1. 1·6 m = 160 cm **2.** 4·4 m = 440 cm **3.** 5·2 m = 520 cm
4. 7·1 m = 710 cm **5.** 3·3 m = 330 cm **6.** 1·9 m = 190 cm
7. 6·1 m = 610 cm **8.** 8·2 m = 820 cm **9.** 3·2 m = 320 cm
10. 2·8 m = 280 cm **11.** 4·0 m = 400 cm **12.** 3·7 m = 370 cm

p 9

cm mm m
m mm mm
m cm m
km mm km

p 10

$\frac{1}{2}$ m 10 cm
5 km 88 mm
300 km 3 mm
100 m 100 m

Now try this!

60 m

p 11

From the top the spikes are: 5 cm, 5·5 cm, 7 cm, 9 cm, 11 cm, 8·5 cm, 8 cm, 7·5 cm, 4·5 cm, 2·5 cm, 1·5 cm.

p 12

1·5 cm 1·5 cm
3·5 cm 3·5 cm
4 cm 4 cm
14 cm
7 cm
8 cm
9 cm
10·5 cm
Check children's stripes are correct length.

Now try this!

66·5 cm

p 13

1. 50 cm **2.** 40 cm **3.** 70 cm
4. 20 cm **5.** 60 cm **6.** 10 cm

Now try this!

1. 50 cm **2.** 60 cm **3.** 30 cm
4. 80 cm **5.** 40 cm **6.** 90 cm

p 15

1. g **2.** kg **3.** kg **4.** g **5.** kg
6. g **7.** kg **8.** kg **9.** g **10.** g

p 16

1. $\frac{1}{2}$ kg = 500 g **2.** $\frac{1}{4}$ kg = 250 g **3.** $\frac{3}{4}$ kg = 750 g
4. $\frac{1}{10}$ kg = 100 g **5.** 2 kg = 2000 g **6.** 4 kg = 4000 g
7. 8 kg = 8000 g **8.** 10 kg = 10 000 g **9.** 20 kg = 20 000 g
10. 1·5 kg = 1500 g **11.** $2\frac{1}{4}$ kg = 2250 g **12.** $1\frac{3}{4}$ kg = 1750 g

Now try this!

heaviest = i lightest = d

p 17

1. 300 g **2.** 600 g **3.** 200 g
4. 100 g **5.** 500 g **6.** 700 g
7. 900 g **8.** 1000 g **9.** 400 g

p 19

1. 180 ml **2.** 110 ml **3.** 90 ml
4. 120 ml **5.** 160 ml **6.** 190 ml
7. 40 ml **8.** 10 ml **9.** 70 ml

Now try this!

8, 7, 9, 3, 2, 4, 5, 1, 6

p 20

3 ml 250 ml
600 ml 1 litre
3500 ml 34 ml

p 21

a 10 ml **b** 700 g **c** 9 kg
d 120 km **e** 140 ml **f** 24 kg
g 45 cm **h** 5 kg **i** 140 ml

Now try this!

mass: b, c, f, h length: d, g capacity: a, e, i

p 23

1. 12 cm **2.** 16 cm **3.** 14 cm **4.** 18 cm
5. 16 cm **6.** 26 cm **7.** 22 cm **8.** 22 cm

p 24

a 6 cm **b** 8 cm **c** 13 cm **d** 17 cm **e** 14 cm
f 10 cm **g** 12 cm **h** 17 cm **i** 15 cm **j** 19 cm

p 25

1. 18 cm^2 **2.** 8 cm^2 **3.** 20 cm^2 **4.** 22 cm^2
5. 20 cm^2 **6.** 10 cm^2 **7.** 14 cm^2 **8.** 13 cm^2

p 26

Now try this!

grey – 6 cm^2 striped – 2 cm^2
black – 2 cm^2 white – 8 cm^2

p 28

1. 31 **2.** 30 **3.** 31 **4.** 31 **5.** 31 **6.** 30
7. 30 **8.** 30 **9.** 28/9 **10.** 31 **11.** 31 **12.** 31
13. 59/60 **14.** 61 **15.** 61 **16.** 62

Now try this!

2 July

p 29

The equivalent durations are:
3. $7\frac{1}{2}$ years
4. 8 years and 4 months
5. approx. 7 years, 8 months and 12 days
6. approx. 9 years, 7 months and 14 days
7. approx. 8 years, 2 months and 18 days
8. approx. 9 years, 3 months and 23 days
9. approx. 9 years, 1 month and 16 days
10. approx. 7 years, 11 months and 3 weeks

p 33

1. 20 mins **2.** 40 mins **3.** 40 mins **4.** 60 mins

10:20 11:25 12:57
10:40 11:45 1:17
11:00 12:05 1:37
11:20 12:25 1:57

Now try this!

Airbus 2 at 9:15 Airbus 4 at 10:40

p 35

Numbers of edges are as follows:

triangular prism	9	rectangular prism	12
pentagonal prism	15	hexagonal prism	18
triangular-based pyramid	6	square-based pyramid	8
pentagonal-based pyramid	10	hexagonal-based pyramid	12

Now try this!

The number of edges of a prism is always three times more than the number of sides of the end face.
The number of edges of a pyramid is always twice the number of sides of the base.

p 36

Now try this!

The cylinder has the fewest faces.
The cube has the most vertices.
The square-based pyramid has exactly eight edges.

p 38

1. 9 **2.** 8 **3.** 10 **4.** 9
5. 6 **6.** 6 **7.** 10 **8.** 12

p 39

1. cube **2.** triangular prism **3.** cuboid
4. square-based pyramid **5.** tetrahedron
6. cylinder **7.** cube

p 40
Zop

Now try this!
Zip

p 42
concave shapes: c, e, f, j, k, m, o	convex shapes: a, b, d, g, h, i, l, n, p

p 43

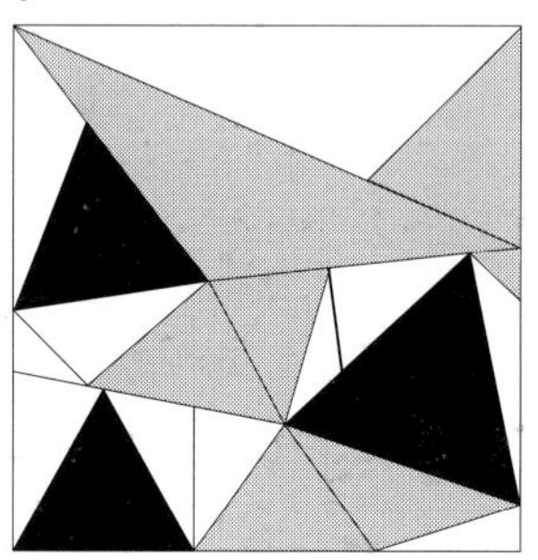

Isosceles triangles are coloured grey.
Equilateral triangles are coloured black.

p 44
regular shapes: a, c, g, i, k
irregular shapes: b, d, e, f, h, j, l, m

Now try this!
equilateral triangle

p 48

shape	0 lines of symmetry	1 line of symmetry	2 lines of symmetry	more than 2 lines of symmetry
a				✔
b				✔
c	✔			
d		✔		
e	✔			
f	✔			
g			✔	
h			✔	
i				✔
j				✔

p 52
IN THE TIN

p 53
1. horizontal **2.** vertical **3.** neither **4.** horizontal
5. neither **6.** vertical **7.** horizontal **8.** neither
9. vertical **10.** neither

p 56
1. NW **2.** SE **3.** E **4.** S
5. SE **6.** N **7.** NE **8.** SW
9. N **10.** NE **11.** W **12.** SW

p 58
1. booby prize **2.** holiday **3.** TV **4.** computer
5. car **6.** watch **7.** phone **8.** bike

p 59

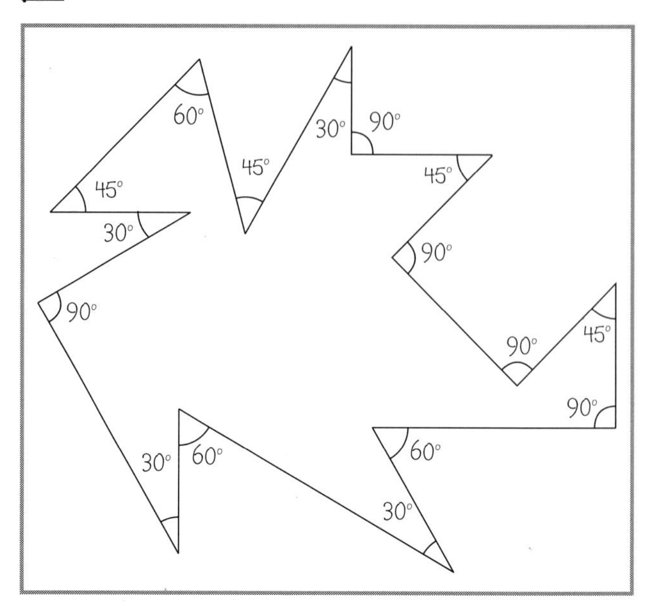

p 60
1. 60° **2.** 120° **3.** 180°
4. 120° **5.** 240° **6.** 240°
7. 240° **8.** 180° **9.** 300°

p 61
1. cat **2.** rat **3.** cow **4.** hare
5. duck **6.** shark **7.** horse

Now try this!
creatures